Herbal Plants
of Jamaica

To Dearest Barbara (Cherry)
and family —

Hope you can have some
fun and raise a few
memories with the book!

With Love

Monica, Aug. 2007

Herbal Plants
of Jamaica
(Bush Teas, Bush Baths, Flavourings and Spices)

Monica F. Warner

MACMILLAN
CARIBBEAN

Macmillan Education
Between Towns Road, Oxford OX4 3PP
A division of Macmillan Publishers Limited
Companies and representatives throughout the world

www.macmillan-caribbean.com
ISBN: 978 1 405065 66 5

Designed by Amanda Easter
Typeset by Carol Hulme
Illustrated by Tek-Art
Cover design by Gary Fielder
Cover photographs by Monica F. Warner

Printed and bound in Thailand

2011 2010 2009 2008 2007
10 9 8 7 6 5 4 3 2 1

Contents

Dedication

This book is dedicated to the precious memory of my mother Miss Rob – Roberta Celeste Williamson, née Jones – who gave each child (Monica, Spencer, Elaine, Errol and Faye) the gifts of life, encouragement and a bottomless well of unconditional love.

Acknowledgements

In writing books of this kind authors invariably draw on many sources of aid and inspiration. The creation of *Herbal Plants of Jamaica* – which was facilitated by Nick Gillard and the team at Macmillan – was no exception.

I am greatly indebted to Kathleen McNary Wood, Dr Sean Carrington and Claire Bordewey (editor) for reviewing and editing the manuscript. I must also pay tributes to the staff and facilities of Hope Gardens, The Institute of Jamaica and the University of The West Indies (UWI), Mona campus; in particular to Dr Trevor Yee of the Natural Products Institute and Mr Patrick Lewis of the Life Sciences Department (UWI).

My gratitude also extends to the Natural History, Bird Life and Geological societies whose thoroughly enjoyable field trips offered perfect opportunities for herb detection. Next, respect to Joy and Sandra and those other Jamaican friends at large who responded to my quest for herbs with pleased smiles, loosened tongues and the treasured gifts of wanted plants.

Finally, loving appreciation to husband George not only for his willing assistance in herb hunting, photography, IT processing and so on, but also for providing that steadying, in-house influence over the entire enterprise.

Disclaimer

This publication is not intended as a guide to herbal self-medication and neither the author nor the publisher can accept responsibility for any such misuse. Persons suffering illness who may wish to use herbal remedies are best advised to consult qualified practitioners.

Introduction

Basket of herbs – *clockwise from top left*: Plump, prickle-edged Single Bible leaves; swatch of small, green Peppermint branches; dry, cream-coloured mass of Irish Moss; dry Orange peel; red-brown Annatto pods; long Fever Grass leaves; lumpy, unripe Noni fruits; coiled, green Cerasee vine and brown, dry Sarsaparilla roots

Jamaicans everywhere will instantly empathise with the contents of the basket above. They will recognise plants – weeds or bush – destined to be made into teas, baths and other concoctions. Herbs like these are mainly wild harvested – gathered from the quick growing forest of greenery known as 'The Bush.'

The following is a much quoted verse or fragment from long ago which itemises the stock-in-trade carried by a female herb seller or 'higgler' on her rounds.

"She had Man Piaba, Woman Piaba
Tom Tom Fall Back and Lemon Grass,
Minnie Root, Gully Root, Granny Backbone,
Dead Man Git Up and Live On turro,
Coolie Bitter and Gorina Bush, the old
Compellance Weed, Sweet Broom,
Cow Tongue, Granny Scratch-Scratch,
Belly-Puller and the Guzu Weed".

Just over half the plants listed in the song above are known to be in use today; those remaining no doubt still function as herbs, but under different names.

What are herbs?

Herbs in the narrow, historical sense usually refer to small, green, short-lived, soft stemmed (herbaceous) plant species used for making medicines as well as for providing food and flavourings in cooking. I have widened this definition of herbs to include all **herbal plants** – whether herbs, shrubs or trees – and those plant parts that contribute to the medicinal, culinary and cosmetic folk practices of the people of Jamaica.

Herbal plants possess elevated concentrations of certain chemical compounds that are either absent or present in much smaller quantities in ordinary plant species. These substances – termed secondary metabolites – seem not to be needed for the primary, day to day activities of herbal plants and are thought to be stored or held in reserve mainly as defence products. Secondary metabolites offer some deterrence – such as being poisonous, distasteful or gummy – against the hordes of microbes, insects, birds and other herbivores that could munch these plants to extinction.

Thousands of secondary metabolites exist. They can be grouped under three main headings: Alkaloids (nitrogen containing substances), Terpenes and Phenolics. Well-known alkaloids include nicotine (from tobacco), morphine (from poppies) and caffeine (from coffee). Terpenes or Terpenoids are defined by their carbon (C) atom content. They are subdivided into monoterpenes, sesquiterpenes, diterpenes, triterpenes and so on whose molecules contain 10, 15, 20 and 30 carbon atoms respectively. Limonene (from citrus) and menthol (from mint) are monoterpenes. Steroids and saponins like the heart drug digitoxin (from the temperate foxglove) are derived from triterpenes but contain 27 or fewer carbon atoms. The red coloured, beneficial carotene (from carrots) is a tetraterpene with 40 carbon atoms. Certain salicylates (from willow) are the phenolic compounds originally processed to make aspirin, while tannins – more complex members of the same group – are common constituents of bark and fruits.

The following is a selection of the remaining types of secondary metabolites. Anthrones and anthraqinones – two closely related polyketide groups that contain some very purgative substances – are present in quantity in the common *Aloe* and *Senna* (*Cassia*) species. Carbohydrates, like starch and mucilage and other sticky semi-solids such as gums and resins, also form part of the huge assemblage of plant protective compounds.

Some of these substances are also varyingly toxic to humans; but they and the others provide the odours, colours and anti-microbial, anti-disease, spasmolytic, nutritive, wound-healing and the host of other useful properties so valued in herbal preparations.

It is hoped that the sparse outline above, together with the information in the Glossary (page 165), will provide some guide to the several chemical compounds that are referred to in the text as well as being a signpost to those readers who wish to pursue the topic further.

Herbal history

Jamaica's history of herbal self help originated with the native Amerindian population and has since been augmented by waves of new arrivals, such as the European colonists, adventurers, African slaves, Indian, Chinese and other workers. This resultant, indigenous herbal healthcare is the first port of call for many ordinary Jamaicans who can neither afford nor easily access the conventional medical system. These and many other citizens believe passionately in bush medicine where the use of plants for their traditional remedies distils down to 'bush teas' and 'bush baths.'

Some Jamaicans have developed a more than ordinary association with herbal matters; with two groups in particular being long-enduring exponents of the use of herbs. First are the Maroons who were historically isolated communities of seventeenth to eighteenth century freedom (from slavery) fighters. These people lived in hiding in remote mountain strongholds and, of necessity, developed a detailed understanding of the flora and fauna of the wild. Their descendants today are represented by the Windward Maroons – in places like Moore Town and Charles Town and villages in the eastern Blue Mountains – and the Leeward Maroons, centred in Accompong, which is a district in the western Cockpit Country. Lately, in keeping with the revitalised national interest in local herbs, Maroon traditional knowledge has been on show in their village gardens, eco-tours, exhibitions and written records. The second group consists of followers of the Rastafari movement. These people – called Rastafarians or Rastas for short – have promoted a liberal, general use of herbs. They also insist that the cultivation and consumption of the 'sacred' or 'wisdom' herb, ganja – the notorious and still illegal marijuana – is a central pillar of their religion.

The island is a recognised hotspot of biodiversity, being home to over 4000 different types of wild plants. About a third of these are endemic, the rest being indigenous or naturalised species. Over time, various parts of an estimated 350 species have been taken into service as herbs. So, buds, flowers, fruits, peels, pods, nuts, seeds, leaves, stems, bark, bulbs, rhizomes and roots may be gathered from trees, shrubs, vines, herbaceous plants, grasses, ferns, mosses, algae, fungi and other botanical groups. These herbal plants and plant parts are each known by at least two common names and are attached to several curative functions or wish lists. Major areas of use are as medicaments, tonics, laxatives and for flavouring in cooking. Common conditions treated include digestive problems, coughs and colds, diabetes, cuts and sores, skin eruptions, women's ailments, fevers, intestinal worms and cancers. Tonics are lightly disguised euphemisms for aphrodisiacs, catering to the already evident male libido.

Processing herbs

Most herbs are minimally processed before use by methods akin to home cooking. Fresh, wilted or dried plant parts are suitably chopped, grated, ground to powder or pounded into fine pieces or a sticky mass. This reduced material can be used directly in poultices (compresses) or extracted in water, alcohol or oil. The commonest treatment of herbs by far consists of lightly boiling the material in water to make infusions or teas.

Poulticing

Plant mash is applied to sore places on the body and secured by wraps or bandages. The material is often warmed before use; oils may be added or the whole thoroughly mixed with cornmeal or other bulky substance.

Extraction in water

Juicing – applied to fresh material. A little cold water is added to the chopped mass and the liquid contents are forced out by squeezing, pressing or wringing.

Infusions or teas – small quantities of herb are put into very hot or boiling water for a few minutes then the solids are strained off; a method often described as 'drawing tea.' The solutions obtained from this and the other processes are the bases for all the teas, beverages, medicines and tonics ingested, as well as the externally used lotions, washes and ointments.

Decoctions – this process is similar to tea-making but is mainly applied to the harder, drier parts of plants such as wood, bark, roots and seeds where the material is boiled for longer periods from 20 minutes up to 2 hours.

Extraction in alcohol

Plant materials are covered with alcohol – usually white rum – and stored in stoppered containers. Alcoholic extracts are known as tinctures.

Extraction in oil

Similar to the extraction in alcohol, except that the liquids commonly used are coconut oil, other cooking oils and castor oil.

Using herbs

As already noted, the overwhelming majority of herbs are used to make water extracts and these are utilised in the main as bush teas, as the decoctions called roots tonics and in bush baths.

Bush teas

These are the products of ordinary tea making and are brewed up to the requirements of dozens of different recipes. They are the herbal remedies and drinks that are the mainstay of Jamaica's ethnomedicine.

Roots tonics

Selection of dried herbs for making roots tonics – *clockwise from top left*: All Man Strength, Birch Gum bark, Blood Wis, Bryal (Bridal) Wis, Chainy Root, Giant Wis, Man Back, Ramoon leaves, Iron Weed, Strongback or Golden Seal, Sarsaparilla Roots, Puron Bark, Milk Wis and Junction Root

These are drinks, beverages, tonics or 'wines' all prepared from extracts of the dried woody, harder parts of certain plants; roots, stems and bark of shrubs, trees and vines (wis) plus added green herbs, nuts, spices, thickeners and sweeteners. Roots beverages, though patronised most by those that are least well off, are recognised by all Jamaicans.

The preparations are expected to put sparkle into tired, worn-out bodies, dredge up stamina, cure many maladies and, most importantly, enhance (male) sexual performance; a process collectively known as 'Good for the

Back'. Consequently, roots tonics – or single ingredients – have been endowed with names such as All Man Strength, Man Back, Cockish Stick, Front End Lifter, Stiff Cock, Keep Me Fit, Stagger Back and other similarly graphic descriptions.

Typical small outlet selling beverages – Westmoreland

Bulk ingredients are available from herb sellers in the markets or from wayside vendors. The drinks themselves – contained in reused plastic or glass bottles – are also offered for sale at these same locations. Smaller quantities of packaged, shredded material are now increasingly found on the shelves of the modern supermarkets.

Wis – also wisp, wiss or wist – are the Jamaican equivalents of the Old English 'withe' where the words refer to the long, flexible bits of a variety of plants. These parts function as herbs but also as rope, whips, wicker and split stems (fibres) for basket making and other crafts.

The plants commonly used to make these tonics come from a selection of about 35 species, not all of which can be unambiguously referred to scientifically described species (see Table 2, page 164). With so many ingredients to choose from, roots herbal tonics vary widely in composition. Up to 22 different plant parts (and more) may be used in any one preparation. Packeted mixtures, intended for small-scale home brews might contain half a dozen core herbs such as Sarsaparilla, Chainy Root, Strong Back, Four Man

Strength, Breadnut Leaf and Medina. For bare essentials the two or three *Smilax* species – Sarsaparilla family – are adequate.

One 14-herb formulation (illustrated) that I bought in a local market contained the following: All Man Strength, Birch Gum Bark, Bryal Wis, Blood Wis, Chainy Root, Giant Wis, Junction Root, Puron (Pruan) Bark, Sarsaparilla Roots, Strong Back, Milk Wis, Man Back and the two green herbs – whole Iron Weed plant and Ramoon leaves.

Birchbark or Tourist Tree trunk

Junction Root plant

Whatever the formula, the basic instructions for making the tonics hold true. The plant materials should be washed, soaked, finely chopped then simmered for between 30 minutes and two and a half hours in generous quantities of water until there has been a large reduction in the liquid. The resulting dark-coloured concentrate is then cooled, sweetened and flavoured

– using cinnamon, cloves, ginger, nutmeg or vanilla – to taste. Molasses is the favoured sweetening agent but raw sugar, honey and condensed milk are also used. Gummarowbit (gum arabic) or Irish Moss may be included as thickeners and alcohol added to strengthen the brew.

Patrons are usually advised to drink one or two (wine) glassfuls of the mixture per day. Commercially produced brands of roots herbal tonics are advertised and marketed at home and abroad.

Bush baths

Bush bath necessities – *clockwise from top left*: Pimento leaves, Barsely leaves and flower heads, heart-shaped foliage of Velvet Vine, Loofahs, Crab Eyes leaflets and giant Cowfoot leaf

Bush baths can be taken in two ways. The first is the traditional, reviving, sauna-like steam bath that is given to persons recovering from illnesses. Here the patient, draped in a covering of blanket or towels, is seated above a tub of fragrant, steaming, herb-strewn water and allowed to perspire freely. The deep sleep that commonly follows this event is said to promote healing.

The second refers to the ordinary, everyday event – immersion in a bath with herbs – that is taken for cleansing, relaxation and skin therapy.

In both cases the herbs used include the more strongly aromatic plants; Barsley, Cerasee, Citrus (Lime), Fever Grass, Guava leaves, Jack-in-the-Bush, Sage, Mint, Pimento and Soursop leaves where a selection of these chopped, boiled leaves – or extracts only – is added to hot bath water. In particular circumstances, the bitter, bright green sudsy extracts that are obtained from Cerasee and Quaco leaves are used exclusively in baths to soothe itchy skin and help heal eczema, rashes, bruises and other eruptions. Loofahs, those bleached, fibrous skeletons of the mature, cucumber-like fruits of the Strainer Vine (*Luffa aegyptiaca*) are traditional body scourers.

There are, of course, many other plants like Ackee, Jointer, Cowfoot, Tamarind, Velvet Bush and Wild Liquorice whose leaves may also be added to bush baths.

Herbs today

Jamaicans continue to source their herbs in the usual ways. Individuals fearlessly pick small quantities of the commonest plants from the roadsides – including the central reservations of busy thoroughfares – and from the hedges, fences and gardens of public and private properties. Other herbs are tended in the yards and fields of persons blessed with some land. But great sackfulls of plants are reaped by professional gatherers in the countryside – from pastures, thickets, mountain slopes and (illegally) from forest reserves. People buy this produce from herb sellers stationed on the street corners or in the open markets. Herbs are also distributed by families and friends and, most importantly, the raw materials, herbal remedies and prescriptions for making home brews are dispensed by traditional healers, herbalists or bush doctors. In some establishments (called balm yards), healing is intricately mixed with superstition, religion and ritual.

Herbs are used singly or in mixtures with some combinations being hallowed by history. For example, there is ginger and lime (or sour orange) with almost everything! There are also cerasee and quaco bush, Irish moss and linseed, sarsaparilla and chainy root. A lot of herb use is opportunistic – use what can be found. This is particularly true of the mixtures used in roots tonics and bush baths. There are also folk guidelines that advise on how many different herbs may be safely boiled together (odd numbers are usual), the best stage at which to gather the plants, and how long certain treatments should last; but these recommendations vary around the island.

Few Jamaican herbal plants have acutely lethal components; many more contain various harmful compounds, but most are safe enough at the dosage levels commonly consumed. Even so, the authorities have had to ban or discourage the public sale of certain plants, *e.g.* the endemic Whiteback (*Senecio discolor*) and Consumption Weed (*Crotolaria fulva*) also called Whiteback because of their proven toxicity (pyrrolizidine alkaloids). Meanwhile, informed sources continue to educate and alert people to the reality of fatalities and chronic illness (liver disease and kidney damage) caused by some herbal practices – particularly in some rural areas where infants and young children are treated daily with bush teas and medicines.

Whiteback flowering shoots showing felty-white undersides of leaves (*S. discolor*)

Honeyed, daisy-flowered inflorescence of Whiteback plant (*S. discolor*)

There is growing concern about the environmental impact and the sustainability of continued indiscriminate, mass collection of wild plants. One response to these problems is the creation of educational, medicinal plant gardens and similar, *in vitro*, germplasm collections. These projects are being spearheaded by the Biotechnology Center at the Mona campus (Kingston) of the University of the West Indies. Decades of research by local scientists into selected Jamaican medicinal plants have produced an invaluable pool of knowledge and some useful spin-offs. Examples include commercial drugs crafted from non-narcotic extracts of marijuana.

Jamaican authorities are in agreement on two goals. The first concerns the need to regularise and improve this complementary herbal health care system (bush medicine) that is used by so many citizens. The second is to increase the national share of the highly lucrative, multi-billion dollar, world neutraceutical industry to which the island already contributes good quality, sought-after products including those from ginger, pimento (all-spice), sarsaparilla and other herbal plants.

I have made no attempt to evaluate the overall effectiveness of the Jamaican use of medicinal herbs except to say that the practice not only endures, it seems to be on its way to new heights of acceptability; as measured by the numbers of different herbs that are packaged and on sale in local and foreign supermarkets. Very many Jamaicans are hugely comfortable with their indigenous herbal legacy and nurture an additional pride in the knowledge that a high proportion of both people and plants have sprung from roots in far away continents.

Arrangement and content

The plants surveyed in this book are separated into major and minor herbs respectively; two categories that are based on the author's elastic perception of the importance of each herb and her own convenience.

I have spoken with local people, tracked down all the plants, grown some in my garden, photographed most and described them all. With this achieved, I hereby present the book to all persons – islanders abroad or at home, visitors and others anywhere who are interested in the living plants behind all those withered bundles, earthy roots, packets of tea, powders, potions, chopped stuff and their often wickedly humorous Jamaican names.

Major herbs

This section contains the bulk of the entries. The major herbs are grouped by family and the families then presented in taxonomic sequence. Therefore, the entries begin with the simplest plant, a marine alga, and progress through the soft stemmed, lily-like herbs to the most advanced species whose families are characterised mainly by woody plants. This arrangement conserves the always interesting relationships between family members. Different herb species within families are usually sorted alphabetically by scientific name.

There are 76 entries from 42 families. Each herb is introduced by its most widely recognised common name with other local names added along the way. The plant is next positively identified by its two-part, Latin, scientific name in italics and if necessary, by its synonym:

e.g. Jamaica Dogwood is   *Piscidia*   *piscipula*   syn. *P. erythrina*
                      (genus)   (species)   (synonym)

Synonyms are the older or redundant scientific names that still remain in use. For simplicity, I have omitted the naming authorities usually attached to scientific names, but these may easily be found elsewhere.

Two or three images of each whole plant or plant parts are included. The plants are characterised to a greater or lesser degree by the following: a brief description; data on origin, habitat and distribution and a synthesis

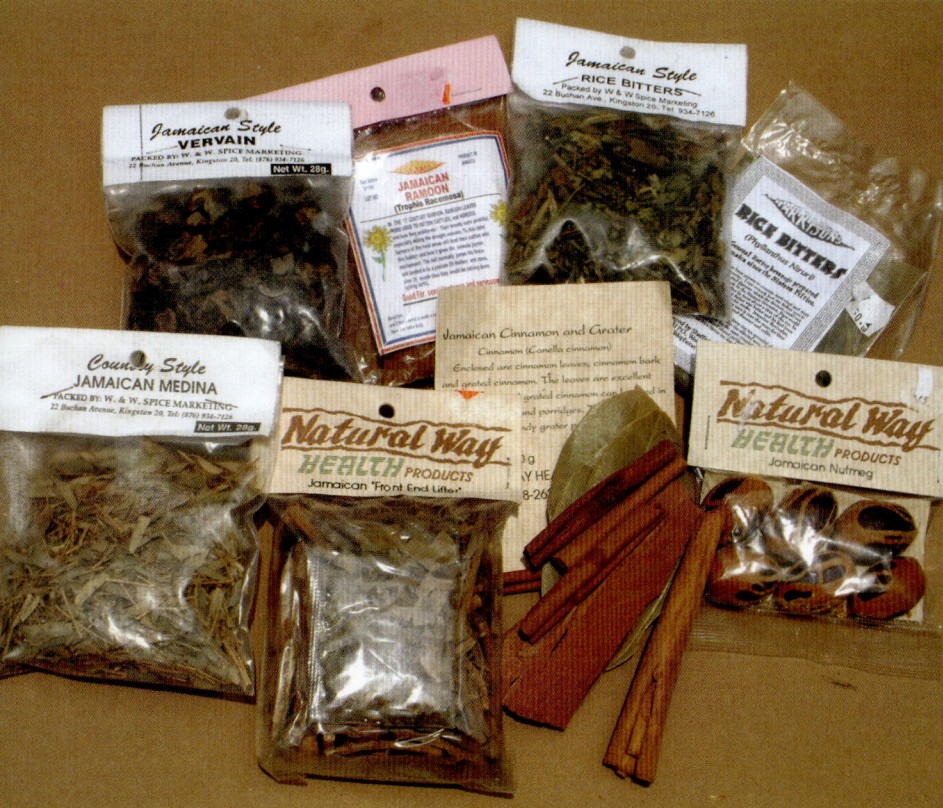

Packaged herbs – *clockwise from top left*: Vervain, Ramoon, Rice Bitters, Rice Bitters, Nutmeg, Cinnamon, Front End Lifter and Medina

of the ethnobotanical use of the species. This last key sector focuses on the following: plant parts chosen; maladies targeted; remedies fashioned; toxicity and the chemical entities – if known – that might be responsible for the herbal properties of the plant. Other items of interest that help to provide a rounded, informative account are also inserted at various points in the narrative. This volume describes and reports only on the general use of certain herbs. It is not prescriptive, there being only two occasions in all its pages where recipe-like instructions are offered.

Minor herbs

This is a short, partly illustrated synopsis of unusual herbal species. Most of the 25 plants are set into interest groups, then these groupings, as well as the single entries, are presented at random. The table at the end of this section lists all the minor herbs by botanical order of family.

Major herbs

RHODOPHYTA – RED ALGAE

♦ **GRACILARIACEAE** *Edible seaweed family*

Irish Moss *(Gracilaria spp.)*
Other names Irish Mash, Punch Back

Live Irish Moss alga

Forget the name: the plant is not Irish, nor is it a moss! It is in fact a red seaweed, a marine alga. In truth, Irish Moss is a mixed bag, referring to *Gracilaria* and several other species of red algae. These plants grow underwater in relatively shallow seas. They make short, bushy plants round about 6 in (15 cm) tall and consist of fleshy, red tinted, branched fronds. These may be flattened like terrestrial leaves, or tubular and sticklike. Red seaweeds, like other marine plants, contain some protein, amino acids, minerals and iodine compounds but they are harvested mainly for their polysaccharide (agar) content. Algal extracts produce agar-agar which is highly valued for its soothing, bulk filling and gelling properties and indispensable for making the agar plates that are used to culture micro-organisms.

Chondrus crispus, the original Irish Moss, is also known as Carragheen. It is a common red alga of temperate waters. This species has long been used in traditional medicine abroad and was the main commercial source of the alternative thickening agents called carrageenans; substances that are very similar to agar-agar.

Irish Moss is also the basis of a very popular, traditional food supplement of the same name and, more extravagantly, is an aphrodisiac in the eyes of many male Jamaicans. The plants can be acquired fresh from fishermen who are said to gather an estimated 80 tons of these edible, red seaweeds annually from Jamaican waters. But Irish Moss is more usually bought in markets and shops as yellowish, dried, sponge-like fragments, nearly always accompanied by small packets of tiny, brown, imported linseed. These seeds are harvested from *Linum usitatissimum*, the blue flowered, temperate plant that also provides flax fibres. When boiled, linseed – or flaxseed – becomes slippery and is a laxative if eaten in quantity. Linseed also contains considerable quantities of oil (with the essential Omega 3 fatty acids), protein and fibre as well as being credited with anti-microbial, anti-cancer and anti-diabetic properties.

Dry Irish Moss fronds with brown Linseed

Preparing Irish Moss starts with boiling the two ingredients together in water for 20 to 30 minutes. The solids are removed and sweeteners, milk, colourings, flavourings, alcohol (rum) and sometimes oats or pureed green pawpaw (papaya) are added to the mucilaginous liquid, which is then left to cool. The mixture will set like a jelly because of the agar extracted from the seaweed. It makes a pleasant tasting and nutritious dish. Irish Moss health and energy drinks are now available in bottles or cans.

Over-collection has diminished the natural supplies of these seaweeds, which has led to the development of farming – particularly in Japan, Florida and parts of the Caribbean – of other red algae such as some species of *Eucheuma* as well as of *Gracilaria*.

♦ ZINGIBERACEAE *Ginger family*

Ginger *(Zingiber officinale)*

Ginger 'roots'

These are the unmistakable, swollen, khaki-colored, underground stems – rhizomes – known as fresh ginger roots. Washed and cleaned, they are sold from heaped piles in all markets. Before use, the pungent, fibrous 'roots' are first peeled, then grated, pounded or chopped. The macerated material is used for making infusions and employed externally in poultices to treat arthritic joints, tight chests, cramps, stomachaches, wounds and other ailments. Ginger teas brewed from the fresh, dry or powdered material are taken primarily to aid digestion and soothe upsets in the system such as colic, diarrhoea, nausea, vomiting or wind. Other uses for the peppery, warming teas include calming headaches, toothaches and treating coughs, colds and influenza. The teas are also traditionally mixed with substances such as honey, milk and rum plus extracts from other herbs like cerasee and fever grass (lemon grass).

Ginger is abundantly used in both savoury and sweet cooking. It contributes to the flavour of the national dish of rice and peas, pickles, fish and meat dishes, beverages and adds spice to baked goods, *e.g.* gingerbread. Grated ginger fermented with sugar, yeast and limejuice produces homemade ginger beer – a firm, old-fashioned favourite.

The main bioactive phytochemicals have been identified as a series of gingerols and a range of terpenoid compounds in the volatile oil that include camphene, neral, geranial and zingiberene. These substances confer the spicy fragrance and fiery taste of ginger 'root' as well as the pharmacological properties of being analgesic, anti-bacterial, anti-fever, anti-fungal, anti-helminthic, anti-inflammatory, anti-spasmodic, anti-tumour and anti-tussive. Other confirmed activities of the extracts include being hypoglycaemic and cholesterol-lowering, as well as stimulating the immune and digestive systems and so protecting against ulcers. All these discoveries support the extensive use of ginger in folk medicine and food.

Pot-grown Ginger plant

Ginger flowers

Mature Ginger flower cone

Ginger is one of the world's oldest and best-known spices. The plant is a native of Asia and although it has been cultivated and exported from Jamaica for over 500 years, only the rhizome is easily recognised. The slender, green stems above ground grow 2–3 feet (66–100 cm) tall and carry narrow, elongated ginger-scented leaves. Gingers are perennial plants. They flower sparingly – at night – during the winter months producing small, cone-shaped inflorescences at the tops of special, short stalks. The cones consist of green, yellow-tipped bracts that sheath the associated buds. In a fortnight or so, when all the exotic, three-part, orchid-like flowers have bloomed and fallen, the bracts ripen to red, after which the entire flowering head gradually fades and decays. At the end of a season's growth, the foliage and stems begin to wither and the rhizomes are harvested. The earliest or young crop is called 'green ginger', while the more mature 'roots' are designated 'dry ginger'. Jamaica's ginger rhizomes, a native yellow and two blue tinted varieties, have long occupied a good international position for quality. Exports of Jamaican dried rhizomes and extractives (oil and oleoresin) are much in demand to service the booming world wide, neutraceutical market.

Turmeric *(Curcuma longa* syn. *C. domestica)*
Other names Tambric, Yellow Ginger

Pot-grown
Turmeric plant

In city supermarkets, Turmeric the herb is visible only as small, curved, banded roots (rhizomes) that are occasionally offered for sale alongside the closely-related ginger. However, the plants are plentiful in some rural areas. Turmeric rhizomes will sprout fairly easily in the garden and grow into 3 ft (1 m) tall, near stemless plants that are really a collection of fairly broad, elongated leaves with long petioles. These plants will acquire new, claw-shaped clusters of rhizomes in a surprisingly short time. Flowering, however, is not as predictable, but the spikes of attractive, tubular, yellow and white flowers are sometimes produced.

Turmeric rhizomes

Turmeric rhizomes whole and cut

The pungent rhizomes are bright orange inside and they are used fresh, dry, raw or boiled. Sliced or mashed rhizome is added, like annatto, to colour and flavour cooking oils and savoury dishes. These ginger-scented morsels are just as likely to leave semi-indelible orange stains on fingers and clothing. Juice from the chewed or pounded fresh rhizome is taken to remedy jaundice, while infusions of many sorts are swallowed to treat liver and digestive malfunctions. All extracts contain sugars, starch and an essential oil with the important yellow phenolic compounds, curcumin and related curcuminoids. Members of all these chemical groups are reputed to work together as anti-inflammatory, anti-microbial, anti-cancer and anti-oxidant agents that also stimulate the immune system and the production of bile. The herb is an established carminative (reduces wind) and cholagogue (encourages the flow of bile).

Turmeric is an ancient Asian (Indian) spice and dye plant that is now grown commercially in many tropical countries. The dried, ground, orange spice is sold around the world and forms the bulk of most curry powders.

♦ **MARANTACEAE** *Prayer-Plant family*

Arrowroot *(Maranta arundinacea)*

Arrowroot plants

Arrowroot is a Brazilian plant that has spread to many other tropical and subtropical places. It is grown commercially in the Caribbean island of St Vincent and sporadically in the cooler, moist districts of upland Jamaica. Their soft herbaceous stems, thin broad leaves and fleshy underground roots (rhizomes) indicate that these plants are members of the ginger subgroup. Small, white, translucent three-part, tubular flowers are produced mainly in June. These beautiful, but transient blooms, may persist for only half a day.

Arrowroot flower

Pot-grown Arrowroot plants

It is the tuberous roots, however, that are important in herbal matters. Stripped of their outer, scale-like, fibrous membranes (bracts), Arrowroot tubers – like those of other *Maranta* species elsewhere – are grated or pounded, then boiled in milk or water to make a porridge-like concoction. The mixture contains a pure form of starch that can be easily digested and therefore is fed to invalids, patients, infants and the elderly. Other preparations are used to counteract the effects of diarrhoea and bowel upsets.

Arrowroot rhizomes

The tubers may also be boiled whole and eaten as crunchy, starchy vegetables. Arrowroot flour has long been available in the shops and is used, like cornflour, as a thickener for sauces, soups and stews. Similarly, both types of flour can be employed as soothing powders or pastes for diseased or inflamed skin. Starchy products obtained from other unrelated plant species may also be given the name Arrowroot.

A slew of over 80 chemicals identified in *M. arudinacea* contain compounds that are known to be anti-leucaemic, anti-PMS, anti-asthmatic, anti-dermatitic, anti-influenza, anti-inflammatory and anti-microbial amongst several other bioactive properties.

The evocative term 'Arrowroot' may have come from the pointed, arrow-like shape of the plant tubers or from the use by Carib warriors of pounded tuber as dressings for wounds caused by poisoned arrows or other mishaps.

♦ **ALOEACEAE** *Aloe family*

Single Bible *(Aloe vera)*
Other names Sinkle or Sinkel Bible, Bitter Aloes, Sempervivum

Single Bible plant

There are about half a dozen *Aloe* species that are used medicinally. *Aloe vera* – a name that translates as the 'true' *Aloe* – is a Mediterranean (North African) species that has had an impeccable history of herbal use from as long ago as the ancient cultures of Eygpt, Greece and Rome. The peculiar 'Single Bible' names used here in Jamaica are probably distortions of 'Sempervivum', which is an unrelated group of Alpine succulents.

Single Bible leaves

The plants grow naturally here in arid coastal areas but are also taken into gardens where they prosper with minimal attention. These unmistakable Agave-like herbs have short, underground stems that continually bud off rosettes of plump, prickly-edged, gel-filled leaves. Flowering occurs over an extended period during the early months of the year when each plant produces a 3–4 ft. (1–1.3 m) tall, branched stalk. The swollen tips expand into wands of attractive, yellow, down-turned, tubular blooms.

Aloe leaves are easily harvested and pulverised into whole leaf or gel only extracts. The pure innermost gel, an essentially colourless and tasteless substance, contains soothing, wound healing polysaccharide and glycoprotein compounds. But the sap close under the leaf surface contains Aloin, which is a bitter, yellow, purgative anthrone. Herbal therapies centre on using the gel or juice as a bitter tonic, an external conditioner for skin and hair and as a cleanser for the internal organs. Fresh gel is eaten for throat and chest complaints, made into hair wash, eye drops and is used as a dressing or salve for damaged skin. Clinical studies have confirmed that Aloe extracts not only hasten the healing of burns, wounds and other injuries, they also reduce scarring. These preparations are also applied to warts, wrinkles, discolorations and other imperfections of the skin. Aloe infusions or drinks are taken primarily as laxatives, but the mucilaginous liquids are also expected to work as abortifacients, treat diabetes and inflammation, heal mouth ulcers, purify the blood and clean both the digestive and urinary systems. Bitter extracts have been historically used to encourage weaning in babies and to discourage finger sucking in children.

There are plantations of *Aloe vera* in parts of tropical America and in some Caribbean islands established to help supply the huge international agribusiness that puts Aloe gel into cosmetics, toiletries, household products and the expanding health products industry.

Single Bible gel

◆ SMILACACEAE *Sarsaparilla family*

Sarsaparilla *(Smilax spp.)*

Sarsaparilla – that nifty roll of syllables meaning little thorny bramble – comes almost unchanged from the Spanish and accurately describes most of the 350 woody, *Smilax* vines. They are a highly similar group of plants that grow in tropical, subtropical and temperate countries and many share the same common name. Using tendrils, their spiny stems can scramble and proliferate into armed thickets or briars. The plants bear leaves that have an odd number of prominent longitudinal veins; some sprout panicles of pale, greenish flowers and, in time, carry the resulting, edible red or dark coloured berries. All *Smilax* species are most prized for their roots though leaves and berries are pressed into service in certain places. From the roots came medicines of the native peoples of Central and South America; remedies for the syphilis and rheumatism of old Europe, and the root beers of American cowboys. Here, they are the indispensable essentials of our local 'roots' tonic industry.

Sarsaparilla vine with
3-veined leaves

Sarsaparilla flowering shoot

Smilax flowers

Jamaican Sarsaparilla (*S. regelii* syn. *S. ornata, S. officinalis*) also known as Honduran Sarsaparilla is one of about half a dozen local *Smilax* species. There is some cultivation of these plants, but in the main it is the slow-growing, wild vines that continue to be heavily over-harvested. Their long, pointed leaves have prickles on the undersides of the three main veins and the berries may be blue, black or red. Below ground there is an abundance of thin, ridged, cream to red coloured, wicker-like roots that are reputed to make the best flavoured extracts.

Pot-grown Sarsaparilla plant

Chainy Root vine and roots

Chainy Root (*S. balbisiana*) is also called China Root or Briar Withe. Apart from having shorter, wider leaves and prickles on the stems only, it is like *S. regelii* in its aerial parts. Chainy Root vines grow up to mid-altitudes and they flower and produce ripe, black fruit for most of the year. The main root of each plant however is a hard, chubby, irregular, red-brown tuber.

Dried *Smilax* roots may be stored for extended periods before being sold off in bundles in the local markets, or in packages in the supermarkets. Decoctions of chopped, dried roots make dark-coloured extracts that are pleasantly scented and have a clean, sweet, spicy flavour. People drink these liquids or teas as treatment for conditions ranging from arthritis, colds, fever, headache, pain, rheumatism, skin infections to weakness and venereal disease. However, Sarsaparilla and Chainy Root are most commonly brewed up with other roots, bark, stems and herbs to produce a variety of herbal tonics. The mixtures are particularly esteemed as male vitality and virility drinks. *Smilax* extracts are mineral rich and are further enhanced by a wide selection of other plant chemicals. These include steroidal saponins (like sarsapogenin) plus organic acids, flavonoids and resin. Compounds like these are reasonably held to be responsible for the anti-inflammatory, anti-psoriasis, blood purifying, diaphoretic, diuretic, stimulant, stomachic and other beneficial functions that are ascribed to species of *Smilax*. Sarsaparilla extracts are in demand commercially for use in drugs, medicinal preparations and as flavouring for beverages and food.

◆ **POACEAE** *Grass family*

Fever Grass *(Cymbopogon citratus* syn. *Andropogon citratus)*
Other name Lemon Grass

Fever Grass plant

Just lightly brushing against the long, narrow leaves of this plant releases a warm aroma of lemon. Rummaging further around in the tussocky waterfall of scented foliage will earn you many itches and scratches inflicted by the stiff, rasping blades of this overgrown grass. Near tubular leaf sheaths branch out into the muted, blue tinged, grey-green leaves themselves. A single 'root' of Fever Grass can grow to a 4 ft (1.3 m) wide composite ball of arching leaves, as each parent stalk buds off a dense stock of new plantlets.

Cymbopogon citratus is an ancient Indian herb that was brought to Jamaica at the very beginning of the nineteenth century and established as a medicinal crop. Today, the plants are still grown in gardens, but wild populations border fences and paths in rural districts. Fever Grass has been historically used here in bush teas and bush baths as a healing herb but only occasionally in cooking. Chopped stems and leaves are brewed up – often with ginger – to make drinks and beverages that are taken as part of the daily diet or used to treat specific complaints such as headaches, tension, nausea, stomach gas and other digestive upsets. The baths, which often include other herbs, are used to break fevers, refresh body and mind and promote restful sleep. Packets of the dried herb are sold in shops.

Fever Grass was named here for its functions in the sickroom but elsewhere the plants are known as Lemon Grass. The main active principle is an essential oil; itself made up of scented, terpenoid compounds. In order of importance they include citral, limonene and linalool, substances that provide that characteristic refreshing, lemony fragrance. Extracts are therefore strongly anti-depressant, anti-microbial, anti-oxidant, carminative, deodorant, insect repellent, tonic, soothing, spasmolytic; effective at healing minor skin irruptions and so well fulfill folk expectations.

Fever Grass stalks

Young Lemon Grass stalks and fresh leaves are very much part of Oriental cuisine. *Cymbopogon citratus* and related species are cultivated in tropical countries for their essential oils. These extracts are widely used in aromatherapy, perfumes, food and drink, cosmetics, soaps, detergents, candles and much else.

Khus Khus *(Vetiveria zizanioides)*

Cropped Khus Khus grass plants at Lyssons Beach, St Thomas

Khus Khus plants were once rather well known here as the source of extracts for making sweetly-scented, but inexpensive, perfumes. These products have now dwindled to a single cologne. The plants are, in fact, rather large grasses from India whose importance worldwide lies in agriculture, perfumery and aromatherapy. Khus Khus leaves are very narrow, keeled (curved) at their bases and angled all the way up; but they are of a beautiful clear lime green colour.

Khus Khus roots – dried pot grown specimen

Fully-grown plants may be 3–6 ft (1–2 m) tall and each supports a network of fibrous roots that can penetrate an equal distance into the soil. This property means that, when planted along contour lines, Khus Khus grass hedges function as meshes that can restrain and stabilise soil and so control erosion. Using this technique, relatively steep slopes may be safely farmed. In addition, the grass does not become invasive as it grows in clumps and its seeds are not fertile.

Fresh Khus Khus plants are not noticeably aromatic. The scent that develops as the grass dries is concentrated in the roots. Bundles of soft, dry roots laid in closets and drawers will deliver a lovely fragrance to garments and linen and will also discourage cockroaches, moths and other insect pests. Small handicraft ventures use Khus Khus leaves and roots as fibres to fashion scented baskets, blinds, mats and other woven items.

All parts of these plants are used in herbal remedies. Infusions of the flowering spike and various plant decoctions are taken as tonics and to ease stomach problems. Root preparations are used for much the same complaints but also to relieve pain, chest ailments and, together with wormseed (*Chenopodium ambrosioides*), to control intestinal infestations.

Abroad, Khus Khus oils (or Vetiver oils) are essential commodities extracted from the dry roots of these plants. They consist of complicated mixtures of mainly sesquiterpene compounds of which vetivone and khusinol are always present. Vetiver oil – described as having woody tones – is strongly aromatic, antiseptic, anti-spasmodic, non-irritant and soothing as well as having fungicidal and insecticidal properties. These oils are major ingredients for the international perfume industry as well as providing popular scents for toiletries, candles, detergents, soaps and other everyday items.

♦ PIPERACEAE *Black Pepper family*

Pepper Elder *(Peperomia pellucida)*

Pepper Elder or Pepperalda is a general, tropical weed that could qualify as the complete herb. The plants appear almost spontaneously in damp, somewhat shady places such as woodlands, flower pots, beneath hedges and along house walls. They are soft, succulent plants that have a spicy smell, peppery taste and carry a raft of other quaint, country names – Man-to-Man, Rabbit Ear, Rat Ear and Ratta-Temper. Their pink-tinged stems grow no taller than 12 ins (30 cm) and carry clusters of appealing, shiny, heart-shaped leaves.

Peperomia plants bear slender, cylindrical flowering spikes – trademark of this tropical family – that are dotted with tiny, round, green or dark coloured (ripe) fruit. This Pepper Elder – together with other close relations of the same name – gathered largely in the moist uplands of St Mary and Portland and pounded with berries of the master spice, Pimento, constituted the basis of the original jerk seasoning. The mixture was used to flavour and preserve the meat of local wild pigs that was then barbequed. Pepper Elder leaves may be eaten raw as a salad or to provide relief for sore throats, while the teas serve to treat many ailments: kidney and eye problems, colds, coughs, fevers, biliousness and the other discomforts of indigestion. The plants are known to have anti-fungal properties and an essential oil that contains common herbal volatiles such as apiole and carophyllene. About 40 other species of *Peperomia* grow in Jamaica, of which several – including the intriguingly named Jackie-mi-Saddle and Tan-pan-Rock – are used as herbs. Many *Peperomia* species are cultivated as houseplants.

Pepper Elder plant in flower

Black Jointer *(Piper amalago var. nigriodum)*
Other names Jointwood, Blackwood or Black Betty

Black Jointer shrub in flower

Black Jointer is an endemic *Piper* shrub. Other shrub members of this genus are similarly large, clearly related plants that are popularly called 'Jointer' or 'Jinta' because of their swollen, joint-like stem nodes. Jointers are also known as and used as Pepper Elder (*Peperomia* spp.) in folk medicine and in soothing herbal baths. These shrubs or small trees grow plentifully about river valleys and are easily recognised by their thin, scented foliage and characteristic, 'mousetail' flowering spikes. The leaves are brewed up – often with ginger – to make teas that are taken as sedatives and as antidotes to fever and stomach problems. Jointer plants provide a plentiful supply of fiery tasting fruiting spikes that make do as black pepper in seasoning savoury dishes.

Black Jointer flowering and fruiting spikes

True Black Pepper plants (*Piper nigrum*) are among the climbing members of the *Piper* genus. They are Asian vines of great antiquity, sparingly cultivated here by small farmers for the berries that are dried to make peppercorns. At maturity, each vine bears dozens of fruiting spikes, so heavily laden that they hang from the slender branches. The vines need stout support. Rural folk once dried and crumbled the mildly aromatic leaves, then sprinkled the fragments as flavouring over food. Mashed green fruits were occasionally employed as stimulating poultices. The condiment, black pepper, is obtained by processing mature green berries, and white pepper is achieved by first removing the skins from ripe fruit.

Unripe Black Pepper berries

Cowfoot *(Lepianthes umbellata, L. peltata* syn. *Pothomorphe umbellata, P. peltata respectively)*
Other name Monkey Hand (USA)

Cowfoot (*Lepianthes peltata*) plant

Cowfoot refers to two native plants that look alike. Both are pale, soft looking shrubs that grow 3–6 ft (1–2 m) tall in damp, somewhat sheltered places. Their prominent, rounded, scented leaves often exceed 12 ins (30 cm) in width. Samples are shredded to make teas that are taken as cold cures. Whole leaves are often swathed around the head to soothe headaches. The plants bear the characteristic, erect but robust flowering and fruiting spikes in the axils of their leaves. *Lepianthes umbellata* – also found in West Africa – is both widely distributed and mildly invasive. The large leaves of this plant are deeply notched at the point of attachment to the petioles or leaf stalks. *Lepianthes peltata* is less common. Its leaves are entire (peltate), their petioles being inserted on the underside, and the plant sap is said to repel ticks. Biologically active compounds reported from this species include anethole (anise flavour) and the pungent piperine that is abundant in black pepper.

Cowfoot (*Lepianthes peltata*) leaf

Cowfoot (*Lepianthes umbellata*) plant

Although some authorities refer both *L. umbellata* and *L. peltata* to a single species (*L. peltata*), from personal observation in Jamaica, I remain convinced that there are two distinct species of *Lepianthes* on the island.

Cowfoot (*Lepianthes umbellata*) flowering and fruiting spikes

♦ **CANNABACEAE** *Hemp family*

Ganja *(Cannabis sativa)*
Other names Cannabis, Collie (Kali) Herb, Marijuana, Sinse (Sinsemilla), Weed

Cannabis – *Cannabis sativa* subsp. *sativa* – flowering plant

Ganja – *Cannabis sativa* subsp. *indica* – young plant

Above is a short selection culled from the 50-odd common names used locally for this most popular of all shrubs. Despite the vast annual harvest here, *Cannabis sativa* – the King of Herbs – remains an illegal commodity in Jamaica and most other countries. The plant is of Asian origin and has been in the human repertoire of useful and medicinal plants for at least 6000 years. Introduced into this country by migrating workers from India, Ganja has become intractably embedded into the island's culture in less than 150 years.

The plants are cultivated with undiminished enthusiasm, usually in small fields tucked far away in the bush. Ganja is a lucrative cash crop for growers: source of the illicit export industry, recreational drug and folk medicine for large sections of the population. Described as annual erect herbs, *C. sativa* plants have square, hollow stems and usually grow 3–6 ft (1–2 m) tall. Some may become trees that are well over 10 ft (3.3 m) in height. The image of their deeply-lobed, serrate, multi-leaflet foliage has become almost an international icon for the species. Ganja shrubs are divided by gender: male and female plants being distinguishable only when in flower. It is the female (pistillate) plants that contain a super abundance of the resinous, mind-altering substance, delta-9-tetrahydrocannabinol (THC) and a range of other cannabinoids. The compounds are phenolic terpenoids. The sticky, THC-rich resin is concentrated on the buds, bracts, flowers and leaves of female shrubs which, if permitted, will mature into intensely aromatic, potent, seeded specimens described as Kali. Unfertilised female flowers have a higher THC content and are known as Sinsemilla (without seeds). Hashish is the resin itself and hash oil is an alcoholic extract of hashish.

Jamaicans consume Ganja in many different ways, but overwhelmingly it is the 'cured', dried, plant material that is used, in the same way as tobacco. Ganja is smoked in homemade cigarettes (reefer, spliff) or in (chillum) pipes for pleasure, pain relief, relaxation, narcotic euphoria and religious (Rastafarian) observance. Medicinally, Ganja has been effective in alleviating the nausea, vomiting and appetite loss created by chemotherapy given to cancer patients. The weed is also regularly used to treat a score of other maladies and injuries. The same rewards are obtained by taking Ganja teas or soups, eating Ganja-spiced baked goods, chewing raw Ganja or drinking Ganja-soaked rum or fruit juice. Informed opinion contends that Marijuana is a dangerous, addictive drug that can be linked to the depression, poor mental and physical health, trafficking and crime displayed by many of the herb's adherents. Additionally, Ganja is accused of being a 'gateway' drug leading to the use of other narcotics …. And so the debate on the legal, medical and pharmacological aspects of this herb continues.

Nevertheless, *Cannabis* plants are a source of over 400 different phytochemicals offering lifetimes of biomedical research. Two local scientists using non-narcotic extracts have developed medicines to treat common, debilitating illnesses. These products include Asmasol, for the relief of asthma, coughs and colds; Canasol and Cantimol eye drops for the treatment of glaucoma and Canavert, used to control motion seasickness.

♦ **MORACEAE** *Breadfruit, Breadnut and Mulberry family*

Ramoon *(Trophis racemosa)*
Other names Raw Moon, Raw Mon

These are trees that may be 60 ft (20 m) tall at maturity and are common to the Caribbean and Central and South America. They grow on low altitude limestone forests here in Jamaica. The plants have a good cover of broad-leaved foliage that is spread over twig-like branches. Inconspicuous male catkins, the female flower spikes and the resulting very small, berry-like fruits are borne on the same tree. Ramoon trees have provided fodder over the centuries, used to supplement the feed of cattle and horses and help stimulate their reproductive urges.

Ramoon bark

Ramoon leaves

Humans also utilise Ramoon trees. Their leaves, bark and roots are often included in the long lists of ingredients that are used for making the aphrodisiac 'roots' tonic beverages. Packets of the dried, powdered, red-brown herb are available in the shops marketed as 'Good for sexual potency and nervousness'. Rural Jamaicans have also used Ramoon 'bush' to help sharpen the eyesight. Interestingly, local scientific research (on animals) has demonstrated that leaf extracts can significantly reduce excessive pressure inside the eyeball, which is one of the symptoms of the common eye disease called glaucoma.

The leaves and plant sap of a close relation, *Artocarpus altilis* (Breadfruit), also count as herbal resources. The seeded form of Breadfruit is one of two Jamaican plants known as Breadnut. *Brosimum alicastrum* is the other and is also used as an herb and fodder tree, the animals being fed both foliage and fruits. The latter may reach 1 in (2.5 cm) in diameter and are sometimes roasted and eaten by country people.

♦ **CHENOPODIACEAE** *Beet and Goosefoot family*

Semi-contract *(Chenopodium ambrosioides)*
Other names Wormwood, Bitter Weed, Bitter Wood, Mexican Tea, Hedge Mustard

These green herbs grow 3–4 ft (1–1.3 m) tall in waste, moisture-rich places. They have ridged, reddish stems and carry a lot of foliage; small pointed leaves that are smooth edged near the apex of the plants but becoming wavy in outline when borne on the lower, older sections. The plants produce long spikes that are crowded with tiny, yellow-green flowers and quantities of even smaller, oily, black seeds. All these plant parts release a strong, definitive and nauseous odour when handled.

Semi-contract plant

Semi-contract fruiting plant with seeds

Herbal Plants of Jamaica

Chenopodium ambrosioides is distributed around the Caribbean and tropical places throughout the world where it is universally used as a vermifuge – an effective folk and medical treatment for worms in humans and livestock. Although a tally of chemicals indicates that the plants are not only anthelmintic and anti-microbial but pesticidal and poisonous, they are much used in ethnic herbal remedies. Extracts such as Oil of Chenopodium made from the leaves and seeds contain the toxic, active principle ascaridole. This is the chemical compound that is effective against intestinal parasites like roundworms (*Ascaris*), hookworms and also against fungal infections such as athlete's foot.

The name Semi-contract (See-mi-contract, Somo-contract) commonly used here in Jamaica is a distortion of the Latin 'semen contra,' which translates as 'seed against'. Jamaicans use Semi-contract against a range of complaints: from arthritis, asthma, colds and fevers to stomach and bowel problems and of course to expel worms, particularly from children. However, over-dosage can be horribly fatal.

Other *Chenopodium* species in Europe and America have been food plants for ages. The shoots, leaves and flowering spikes of *C. bonus-henricus* (Good King Henry) and *C. album* (Fat Hen) are eaten as spinach. Native Americans used to grind the abundant seeds of Fat Hen into nutritious flour. But *C. ambroisiodes*, also called American Wormseed, is present in the lists of poisonous Jamaican plants.

Semi-contract young shoot

♦ AMARANTHACEAE *Callaloo family*

Devil's Horse-Whip *(Achyranthes aspera var. aspera)*
Other names Hug-me-close, Colic Weed

Devil's Horse-Whip plants with long spiny spikes

This widely distributed, tropical weed can grow to about 3 ft (1 m) tall in sun and twice that height in shady situations. The scrambling herbs have pairs of soft, rounded leaves and are remarkable for sprouting many spiky, whip-like inflorescences. They may extend up to 18 ins (45 cm) in length and develop a succession of tiny, bead-like, whitish blooms. Eventually, the thin, ridged flower stems become studded with small, scale-covered fruit that are like hazards when mature. These fruit, by then brown and prickly, will dislodge if carelessly touched, impaling themselves painfully in skin and annoyingly on clothing, hair and fur. As vegetable scourges, these fruiting spikes might well be effective on the horses of the Devil – presuming that the terrified beasts needed any extra urgings.

In Jamaica and a few other Caribbean islands, brews made from the leaves, whole plants or a combination of leaves and roots are used to treat a range of complaints. Chief among these are coughs, colds, colic, chest pains, fever, influenza and venereal disease. Children sick with colds or malnourishment are also treated with these remedies. Some plant infusions are regarded as tonics and aphrodisiacs.

Achryanthes aspera, rare in Jamaica, is an altogether larger plant much studied abroad. This species has had a long history of use in China and India where it seems that all parts of the plant are employed in making herbal preparations for treating multiple complaints.

Betaine, the alkaloid achyranthine, the steroid ecdysterone, sugars, saponins and a high mineral content (potassium) have been identified from plant and seed extracts of *A. aspera* and *A. bidentata*. These and other substances provide the anti-inflammatory, anti-fever, analgesic, protein-building, diuretic, cardiotonic, astringent and other properties of these herbs. It seems likely that the Devil's Horse-Whip, *A. aspera var. aspera*, shares some of these characteristics.

Devil's Horse-Whip tip of flowering spike

♦ **PHYTOLACCACEAE** *Jocato and Pokeweed family*

Dogblood *(Rivina humilis)*
Other names Inflammation Weed, Dogberry, Cat's Blood, Pigeon Berry

Dogblood shoot

A common annual, tropical plant of the lower altitudes, Dogblood is almost always found in the shady, weedy edges of paths, fields, forests and open places. The shrubs branch freely but grow no taller than 3 ft (1 m). Though bearing soft, pointed, green leaves, the plants are usually highly visible because of the attractive fruit and flowers that are present throughout the year. Stalks of tiny, pale pink or white blossoms give rise to bright red berries that were once a source of colouring or dye for skin and fabric. Birds are quick to feed on the poisonous fruits but they suffer no ill effects. The crushed berries make blood-coloured children's play paint! Rouge plant, yet another descriptive name for this herb, is mostly used abroad.

Dogblood flowers

Dogblood berries

Dogblood extracts are taken to control inflammation, but the liquids are also generally used as remedies for colds, diarrhoea and marasmus. In parts of Portland, where Dogblood is additionally known as Belly Bush and Period Pain Bush, herbal teas and baths are employed to address menstrual problems. Infusions made from the entire plant, berries, leaves or roots are used in parts of the Caribbean to treat cancer, fever, jaundice, stomachache, wounds and other ailments.

Family member *Phytolacca rivinoides* – the wild Jocato – is commonly boiled and eaten as spinach, though the plants are said to contain some toxic principles. Both Dogblood and Jocato are clearly similar to the related American pokeweed *Phytolacca americana*. Young Pokeweed shoots (and roots) are eaten as cooked greens and all plant parts are used medicinally in America. But, despite the very long history of indigenous folk use, Pokeweed herbs still occasionally cause serious illnesses and even fatalities – possibly because of toxic plant lectins and saponins.

White-flowered Jocato plant

Pink flowered Jocato with red stems

Guinea Hen Weed *(Petiveria alliacea)*
Other names Duppy Weed, Strong Man's Weed, Gully Root, Stink Weed

Guinea Hen Weed in flower

Guinea Hen Weeds abound at lower altitudes in the nooks and crannies of gardens, alongside fences and roadsides, in pasturelands and open waste places. Most favoured habitats for this persistent, perennial, pan tropical species are in semi-shaded undergrowth where the plants can proliferate to form chest-high thickets. They have shiny, wavy-edged leaves and bear tiny, white, starry blooms dotted along one side of a long, thin, arching flowering spike. As the flowers fade and the small, dry, hooked fruits develop, so the spike straightens until it is finally erect. These fruits have a high nuisance value, as they easily embed themselves into skin, fur, hair and clothing. The slender stems and deep-questing white roots are tough, sinewy growths that resist breakage. All parts – roots in particular – emit a strong 'stink' of garlic when the plants are trampled, bruised or damaged in some way. This taint can be transferred to the flesh and milk of cattle that have fed on the plants.

Foliage and small whole plants are laid in bedclothes, or crumpled and the pungent odour inhaled in efforts to banish fever, headaches, duppies (ghosts) and related supernatural problems, while a mixture of root in rum is a well-known remedy for toothache and for filling tooth cavities. As an aid to freshwater fishing, crushed plant parts used to be thrown into rivers and the resulting stunned, floating fish easily netted downstream. Decoctions, teas and herb mixtures are taken as sedatives and also used to treat bladder inflammation, diarrhoea, rheumatic pain, skin diseases and many other ailments. The plants are used in similar ways throughout the Caribbean. In tropical Central and South America, *P. alliacea* – known as Anamu – has an extensive, historical herbal reputation with over 60 known applications that range from abortifacient to vermifuge. Scientists here and abroad have confirmed that dibenzyltrisulphide – one of the many bioactive plant chemicals in Guinea Hen Weed – is effective in boosting the human immune system, and may be translated into a future cancer cure. The compound also kills insects and ticks.

◆ **CACTACEAE** *Cactus family*

Tuna *(Opuntia tuna)*
Other names Toona, Prickly Pear, Roast Pork

Tuna plant

Four other species – Cochineal Cactus (*O. cochenillifera*, *O. ficus-indica*), Seaside Tuna (*O. stricta var. dillenii*) and the endemic *O. jamaicensis* – no local name – are commonly called Tuna and used interchangeably with *O. tuna*. *Opuntia* species are distinctive, branched cacti made up of fleshy, flattened pads or stems that may or may not be spiny. These stems have some uses in common with the gel-filled leaves of Sinkle Bible or *Aloe vera*. The skins and spines of Tuna stems are removed by peeling or roasting and the succulent flesh sliced or pulped for various uses. The resulting slippery, soapy mash was once widely used as a rural shampoo to clean and condition the hair as well as to protect against dandruff and related ills. Sliced Tuna flesh is applied to cuts and burns and, together with salt, used as poultices for sore and damaged skin. For internal use, the mucilaginous cold-water extracts are taken to treat fevers and constipation and tuna teas drunk to relieve menstrual problems. These plants are also used medicinally in tropical America and other Caribbean countries.

Tuna flower – short petals and long pink stamens clustered around the yellow claw-like stigma

Opuntia cacti flower brightly in red, yellow or orange and their elongated or pear shaped fruits are similarly coloured, sweet and edible. The roasted flesh of these plants – which was once consumed by the neediest folk here in Jamaica – is said to resemble that of cooked pork. Elsewhere, Cochineal Cactus and other cacti are cultivated commercially for both fruit and stems which are used as vegetables.

Soursop flowers

◆ **ANNONACEAE** *Custard Apple family*

Soursop *(Annona muricata)*

The yards and gardens surrounding most Jamaican dwellings have space enough for at least one of these tropical American trees that are cultivated for their scented fruit and foliage. These are tough, wiry plants growing to about 20 ft (6.7 m) tall, able to endure injury and neglect, while still producing the large, prized, irregularly-shaped fruit known as soursops. Their rough, brittle skins remain green even when the fruit is ripe. Inside is a fragrant mass of sourish, white pulp and a host of large, black, cyanide-containing toxic seeds, which must be removed before the fruit is consumed. Although the creamy flesh is delicious eaten raw, it is normally made into a pleasant, cooling beverage that is flavoured with lime juice and nutmeg and sweetened with sugar or condensed milk. The fruit pulp is also used in creating ice creams, sorbets and other frozen delights.

Soursop fruit

Soursop, in any form, is valued as a soporific: a soothing calming agent. It is held to be 'good for the nerves,' an anti-spasmodic and the suggestively shaped heart, or core, of the fruit is said to be particularly effective when eaten. Soursop leaf teas (and juice) are taken for hypertension, urinary problems (bedwetting), worms and parasites. Other common applications are as antidotes to poisoning, fever, colds and coughs. The leaves are added to bush baths. Soursop bark and roots also feature in herbal preparations. Elsewhere – in Tropical America and the West Indies – the crushed seeds are used as insecticides and emetics as well as in treating a wide range of complaints including those listed above. Recent research has indicated that Soursop extracts are a potent force against many types of cancer cells.

At least 50 phytochemicals have been identified in Soursop extracts. They include alkaloids, terpenoids, phenolics, tannins, organic acids, volatile oils and vitamins which, together with other substances, confer the analgesic, anti-cancer, anti-fever, anti-parasitic, anti-spasmodic, diuretic, nerve-soothing and feel-good actions among many other valued qualities.

The properties and characteristics of Soursop are present to some extent in all *Annona* species, of which there are several.

Annona squamosa is the popular fruit snack called **Sweetsop** or Sugar Apple. The very sweet, much praised, Andean Cherimoya (*A. cherimola*) and the smooth, grey-skinned **Custard Apple** or Bullock's Heart (*A. reticulata*) are both more seasonal and less plentiful.

Sweetsop flower

Custard Apple flower

Sweetsop fruit

Custard Apple fruit

The fruits of various other wild-growing members of the genus, though edible, may not be as available, juicy or palatable as the ones described above. They include the Pond or Alligator Apple (*A. glabra*), the hardy Mountain Soursop (*A. montana*) and the endemic, Wild Soursop (*A. praetermissa*).

♦ **LAURACEAE** *Advocado Pear family*

Cinnamon *(Cinnamomum verum (syn. C. zeylanicum))*

Young Cinnamon tree

Jamaicans are very familiar with cinnamon – sometimes called cinnamon leaf – which is available as a pungent red-brown powder or as selections of recognisable plant parts. The latter are sold in packets containing brittle, dry olive-green leaves, curls or chunks of bark – called sticks – and at times accompanied by small, locally crafted graters and home-grown nutmegs. The leathery, fresh leaves are also in demand. They are of a light green colour, each featuring a prominent trio of veins.

Cinnamon leaves

Cinnamon – fresh and dried leaves plus curls and sticks of bark

Few realise that cinnamon is a herb tree as well as being a source of spice. The plant is native to India and South Asia, regularly 30 ft (10 m) tall at maturity and, particularly when young, provides the scented bark, leaves, twigs and roots used in the centuries-old spice trade. Cinnamon is a very ancient commodity, now grown commercially in many different tropical places.

Cinnamon the spice is used to flavour baked goods – bread, biscuits, buns, cakes – desserts, puddings, other sweet dishes and drinks. Cinnamon in porridge has been a favourite for generations.

Herbal infusions made from the powdered materials – mainly bark – are popular choices for treating arthritis (poultices), digestive problems and menstrual disorders. These teas are also taken as general stimulants. Leaf teas – one leaf per cup – provide pleasant daily beverages.

The extracts mentioned above contain very small amounts of fragrant, potent plant oils that include cinnamaldehyde, eugenol (oil of cloves), linalool and camphor from the root bark. Substances like these are analgesic, stimulant and also strongly anti-microbial. They kill the bacteria, fungi and viruses that cause common malaises such as mouth and gum ulcers, stomach upsets, diarrhoea, food poisoning, infections of the urinary tract and thrush. However, the purified and concentrated herb oils should not be used. They are toxic and known to cause allergies, dermatitis, liver and kidney damage amongst other reactions. Pregnant women and nursing mothers are advised not to use cinnamon medicinally.

The closely related *Cinnamomum camphora* or Japanese Camphor Tree was also introduced here. This species is the original source of commercial camphor.

Wild Cinnamon or Canella (*Canella winterana*) is an indigenous shrub whose aromatic bark contains much the same essential oils as the *Cinnamomum* species and is used in similar herbal therapies elsewhere. Found often in dry, coastal thickets, the plants have leathery leaves and their fruiting shoots bear small, attractive, bright red flowers and decorative bunches of white and red (ripe) berries.

Wild Cinnamon flowers

Wild Cinnamon fruit

♦ **MENISPERMACEAE** *Moonseed family*

Velvet Leaf *(Cissampelos pareira)*
Other name Pareira Brava

Velvet Leaf vine

Fuzzy, young, grey-green vines of Velvet Leaf sometimes twine through the undergrowth, but they are best seen trailing on the sunny sides of fences, thickets and trees in woodland edges and clearings. These common, tropical plants are clothed in soft, rounded hairy leaves that range from 1–5 ins (2–12 cm) in width. Male flowers hang in airy, branched panicles leaving the more compact, less conspicuous female flowers to produce tiny, red, berry-like fruits for most of the year.

Velvet Leaf leaves

Velvet Leaf flowers

The inviting, velvety, light green foliage and young stems are used in bush baths and brewed into diuretic infusions or teas for asthma, bowel problems, rheumatism and other complaints. Freshly bruised leaves are applied to animal bites, stings, cuts and wounds.

Jamaicans make modest use of Velvet Leaf compared with the comprehensive, centuries-old, medicinal applications of some indigenous South American people. Dubbed the 'midwife's herb,' plant preparations have been particularly linked to the treatment of female complaints. Extracts are employed to mediate the powerful contractions of the muscular uterus during childbirth, cramps, haemorrhages, difficult menstruation, miscarriages and other conditions. Medications based on this plant are on sale in the Americas.

Nearly 40 alkaloid phytochemicals have been identified in leaf and root extracts of Velvet Leaf. They include substances like tetandrine that are responsible for the analgesic, anti-bacterial, anti-fever, anti-inflammatory, anti-spasmodic, anti-haemorrhagic, muscle relaxant, hypotensive and other properties of this plant.

Cissampeline – a drug that relaxes skeletal body muscle – was developed from Velvet Leaf extracts. However, as overdosing on home-made remedies can result in sedation, heart irregularities and eventual heart failure, *Cissampelos pareira* is included in lists of poisonous Jamaican plants (and was one of several plants used in making South American arrow poisons).

◆ **PAPAVERACEAE** *Poppy family*

Mexican Poppy *(Argemone mexicana)*
Other names Prickly Poppy, Yellow Thistle, Spirit Weed, Tissel

Mexican Poppy plant with spiny pods

Mexican Poppy is a common, prickly, tropical roadside weed occasionally seen here in small, cultivated patches. The plants are generally below 2 ft (60 cm) tall with stiff, lobed, spiny leaves. Each bright yellow, cup-shaped flower has a red spot in the centre composed of stigmas. When dry, the fruit or capsule, splits into several segments releasing numerous, tiny, round, black seeds. These are said to be narcotic and can be used to provide oil. Plants that are broken or bruised ooze an acrid, yellow sap that is applied to skin ailments such as boils, wounds, sores, ringworm and warts.

All parts of these plants are poisonous, being generously supplied with potent isoquinoline alkaloids; they include sanguinarine (seeds and roots), berberine (leaves, seeds and roots) and protopine, which occurs throughout all the tissues. Nevertheless, *Argemone mexicana* is a much used, highly popular herb throughout Central and South America, parts of the Caribbean and Ghana in Africa.

The bush teas, which are known to have pain relieving and calming properties, are used here for a good round of complaints, some of which are mentioned below. Infusions made with the flowers are taken to banish insomnia and soothe teething children. Leaf juice is applied to inflamed eyes while leaf juices and teas are taken for asthma, colds, fever and high blood pressure. The stronger plant decoctions and seeds are purgative and used to treat disorders of the bowels and urinary system and epilepsy. The seeds and leaves will function as insecticides.

Mexican Poppy is recognised as one of the poisonous plants of Jamaica, as two of the compounds listed above, if taken in quantity, can affect the central nervous system, the respiratory system and the muscles of the heart and uterus.

♦ **CRASSULACEAE** *Stonecrop family*

Leaf-of-Life *(Kalanchoe pinnata* syn. *Bryophyllum pinnatum)*
Other names Love Bush, Live Forever, Life Plant

Leaf-of-Life flowers

Kalanchoe pinnata is a Madagascan species now thoroughly integrated into the natural vegetation and herb gardens of this country. They are distinctive, succulent, 3 ft (1 m)-tall plants that thrive up to an altitude of around 4000 ft (1333 m). Leaf-of-Life plants are at their most dramatic during December to May when they produce bunches of drooping tubular flowers atop lily-like stalks. Perversely, those plants growing in the least favourable places – crumbling, arid hillsides – seem to bear the brightest blooms. Young purplish stems bear the fleshy, paired, lobed to compound leaves. Older stalks are distinctly ringed. Each pale green leaf has scalloped, purple edges that contain a series of buds. Therefore, new plantlets are able to materialise from the notched margins of fallen (or attached) leaves giving these herbs a reputation of constant rebirth and rejuvenation; hence the basis of many local names and extensive medicinal use of the herb.

Leaf-of-Life leafy shoot

Kalanchoe leaves and young stems are chopped, juiced or eaten raw, often with salt. This combination is felt to be particularly good for general well-being, asthma, bronchitis, colds, the gut and hypertension. Leaf-of-Life teas are also taken for the ailments listed above as well as for kidney and menstrual malfunctions. Externally, warmed leaves, crushed leaves or leaf juice are applied directly to boils, bruises, cuts, eye and ear disorders, headaches, insect bites, swellings, skin and mouth ulcers. Research has confirmed that these soothing, mucilaginous and healing plant extracts are anti-bacterial, anti-fungal and anti-inflammatory. Leaf-of-Life extracts, like those of other plants, contain organic acids, phenols, flavonoids and minerals. This herb is sometimes used in combination with the leaves of other plants such as Seaside Mahoe (*Thespesia populnea*) and Yam vines (*Dioscorea* spp.). In tropical South America, *K. pinnata* – popularly called Miracle Leaf – has extensive medicinal uses.

Leaf-of-Life plantlets growing from the edge of a leaf

Herbal Plants of Jamaica

◆ **CAESALPINIACEAE** *Senna family*

King-of-the-Forest *(Senna alata* syn. *Cassia alata)*
Other names Candlestick, Ringworm Shrub

King-of-the-Forest flowering at the edge of a gully in Kingston

Originally from Tropical America, this member of the prolific *Senna* genus (many of which were once classified in the *Cassia* genus) has spread to hot countries around the world. The plants grow naturally here but they are also cultivated. They thrive best in moist or swampy places from sea level up to 3000 ft (1000 m). The distinctive, pinnate foliage, which consists of large, glossy leaflets, is arranged more or less horizontally but the plants become positively eye-catching when in flower. From October to April long flowering stalks bear big sculptured, flame-yellow buds. These mature successively, first shedding their orange bracteoles then opening yellow petals. The resulting pods, which contain a large number of seeds, become black when ripe.

King-of-the-Forest flower head

Fresh leaves are a sought after herbal resource that is sold in the folk markets. Leaf infusions are taken to treat 'pressure,' blood complaints, coughs and colds, and pounded leaves and plant sap are applied to skin eruptions. Analyses of plant extracts have identified anti-microbials, fungicides, laxative anthraquinones, saponins and other chemical compounds that are effective at treating skin diseases like ringworm and expelling intestinal worms. Candlestick and other *Senna* species are also known as Cassias – plants whose leaves and pods are renowned for their purgative properties and abilities to heal skin infections.

Overdosing on plant extracts can cause seriously toxic side effects. These include loss of weight through lack of appetite, nerve damage and eventual death.

Dandelion *(Senna occidentalis* syn. *Cassia occidentalis)*
Other names Piss-a-bed, Wild Coffee, Stinking Weed

Dandelion plant with flowers and ripe pods

These are rather sparse, tropical shrubs – 3–5 ft (1–1.5 m) tall – that normally grow in small groups along roadsides and in waste places. However, given a change in conditions, like the occurrence of storms and heavy rainfall the colonies often expand into substantial thickets. The plants have four to six pairs of red-stalked, red-rimmed and red-veined leaflets that have a faintly unpleasant smell when bruised. Like other *Senna* and *Cassia* species this herb has a reputation for being mildly purgative but effective at treating skin disorders.

Dandelion shrubs bear a succession of small, yellow, down-turned blossoms. They are the precursors of the crops of narrow pods each of which contains an impressive cargo of tiny, olive-grey seeds. These are gathered, parched (roasted), and then ground to a powder for making coffee-like drinks. The drinks are indeed taken as beverages but also commonly given to children as an aid to control bedwetting (piss-a-bed) and related bladder and kidney problems. Overdosing on these or other preparations can prove to be strongly diuretic and may also cause severe stomach upset. Similarly, over-consumption of the pods and seeds can be fatal to grazing animals. Pregnant women are advised not to consume Dandelion products.

Dandelion shoot, pods and seeds

Crushed leaves, plant juice, whole plant decoctions and extracts of leaves, flowers and roots are employed to treat many illnesses. These include bronchitis, fever, headache, infection, menstrual and uterine pain, inflammation, intestinal parasites, skin troubles and urinary and liver complaints. Dandelion plants contain anthraquinones, flavonoids , sitosterol and other active compounds that account for both the medicinal and toxic properties of this herb.

Senna occidentalis is not related to the European dandelion (and North American weed) *Taraxacum officinale* which is a completely different plant, a member of the Compositae family.

♦ **FABACEAE** *Pea family (Leguminosae)*

John Crow Bead Vine *(Abrus precatorius)*
Other names Crab Eyes, Liguish, Red Bead Vine, Wild Liquorice, Rosary Pea

This widespread bush plant grows particularly well in dryish areas that are less than 1000 ft (333 m) in altitude. Mature specimens consist of lush green foliage and a multitude of thin, woody, branching stems that twine over other shrubs, trees, fences or anything else that provides support.

Abrus precatorius is an attractive but potentially very dangerous plant that has spread from Southeast Asia to most warm regions of the world. The seeds are the famous Jequirity beans of India that are used as very small units of weight, each 1–2 grains (65–130 mg), as well as being a source of domestic

Crab Eyes flowers and unripe pods

poison! Here in Jamaica, the vines, which have acquired a mixed string of factual and fanciful local names, are listed amongst the 50 most poisonous common plants in the island.

Crab Eyes bunches of ripe pods and seeds

John Crow Bead Vines bear long, graceful, alternate leaves each composed of 10 to 20 pairs of small leaflets. When vigorously crushed between the fingers these leaves provide a faint, sweetish smell of liquorice! Extracts from boiled down leaves and roots are used to fabricate medicine for coughs, while stems and roots have provided alternative sources of liquorice or 'liguish'. This substance is a well-known ingredient of commercial cough medicines. It helps to remove mucus from the respiratory passages. True liquorice – that challenging, dark, treacle-like substance made from the roots of a long known, fellow pea plant *Glycyrrhiza glabra* – is commonly available in sweets of the same name in the UK. Here, Liguish teas are also used to treat at least half a dozen other complaints including asthma, fever, headache and heart problems. Some preparations function as emetics and – like much else – aphrodisiacs!

A tangle of small pods follows each cluster of pretty, purplish-pink pea flowers. The mass of curled, twisted pods split when ripe, fully exposing their shiny, bright red seeds, each bearing a black 'eyespot'. These inviting, hard-shelled seeds were once much used in craft (beadwork). They were strung into necklaces, rosaries and belts, used in ornaments, and as the 'eyes' of toys, and were also enclosed as the rattling seeds of maracas! However, some of these practices have been discontinued because of the poisonous nature of the raw material. Unbroken, Crab Eyes seeds are safe and have been swallowed with no ill effects. But, if they are chipped, pierced or chewed, then the deadly toxin Abrin, may enter the bloodstream via any small cuts or wounds or from the gut causing severe illness or death to humans and livestock. One chewed seed is more than enough to fell an adult. It is known that cooking (boiling) or the action of the digestive juices destroys the poison.

Nevertheless, Abrin (a lectin) and some of the many other phytochemicals present in this legume are being screened for use against diseases such as HIV, cancer and arthritis.

Crab Eyes mature pod and seeds

Medina *(Alysicarpus vaginalis)*

Medina plant with trailing stems

People often stop by the waysides to gather and bundle the thin trailing stems of Medina with an eye to brewing tasty – and hopefully aphrodisiac – teas and beverages. The weed is universal in lawns and other places where the vegetation is periodically close cropped. The tough, flexible plants easily adapt to growing horizontally as well as blending in with clover and the running stems of grass.

Packets of the chopped, dried herb are now sold commercially with instructions for making a tea promised to alleviate nervousness and bladder weakness when taken twice weekly! Medina is an Asiatic plant now naturalised in tropical and subtropical places such as Florida and the West Indies. Known

as Alyce or Buffalo Clover, *A. vaginalis* is cultivated abroad as nutritious, fresh fodder and hay. It is also used locally as a feed to stimulate racehorses. Grown as a crop, the plants remain erect up to 3 ft (1 m) tall and re-seed vigorously. Medina plants anywhere bear simple, alternate leaves, pretty reddish, pea-type flowers and the resulting clusters of tiny pods.

Medina flowers and pods

Jamaica Dogwood *(Piscidia piscipula* syn. *P. erythrina)*
Other names Fishfuddle, Fish Poison Bark

Dogwood bark

Native to Tropical America and the West Indies, Dogwoods are common in the thin, dry, coastal soils and limestone forests of Jamaica where the deciduous trees may exceed 60 ft (20 m) at maturity. Both generic – *Piscidia* – and local names reflect the widespread indigenous fishing practice of putting crushed plant parts into the water to stun or stupefy fish. This occurs because the chemical compound rotenone (or derris), present in Dogwood interferes spectacularly with the uptake of oxygen by fish and invertebrates (mites, spiders, insects and larvae). Rotenone, an isoflavone, is widely used in commercial pesticides. The remaining constituents in Dogwood include organic acids like piscidic acid, other rotenoids, tannins and saponins.

Extracts of Dogwood bark have for long been included in traditional remedies for the general relief of pain and muscle spasms and in drugs that are used as sedatives. Therefore, the medicines are taken for coughs, headaches, insomnia, nervous tension, neuralgia, painful menstruation, toothache and related ailments. These extracts also act against fever and inflammation. Children, old people, pregnant women and nursing mothers are advised not to use the herb as over-dosage and inappropriate use can cause convulsions and other serious side effects.

Dogwood buds, flowers and leaves

Immature Dogwood pods

Dogwood shrubs and trees – like other family members – bloom beautifully on bare branches towards the end of spring. Short, conical, inflorescences bear a multitude of small pea flowers, which have flaring white petals, yellow centres and dark, toothed calyces. The resulting bunches of young fruit are short green pods, each defined by four, projecting longitudinal rows of papery wings.

◆ **RUTACEAE** *citrus family*

Lime *(Citrus aurantiifolia)*

Lime flower and tiny fruits

Lime trees – of all the citrus species – are the commonest, most versatile and deeply appreciated items in the Jamaican yard. The trees, originally from Asia, make handsome specimens covered in dark green, finely toothed, aromatic foliage. They bear bunches of sweetly-scented, waxy-white flowers year round that mature into clusters of the familiar rounded, green skinned fruit called limes. Lime products are applied almost seamlessly over the culinary, domestic and medicinal fronts.

Grated fruit peel (zest) is used to spice both sweet and savoury dishes. The acid juice – from ripe (yellow) or green fruit – makes popular cooling beverages, flavours other fruit juices and is an essential ingredient of our national rum punch. Limejuice – plus salt – makes an effective, anti-bacterial scrub that removes stains from skin, fabric and hard surfaces. The solutions 'cut' grease and deodorise the unpleasant smells that often emanate from fish, meats and sweat. Fruit juice, as well as the infusions of young leaves, stems, buds, flowers and peel (dried) make refreshing drinks that are helpful in settling upset stomachs, countering nausea, colds, fever, headache, toothache, influenza, sore throats and other common ills. These same plant parts can transform baths into relaxing, therapeutic experiences. Lime extracts make good home disinfectants (washes) for abrasions, cuts, sores and minor skin infections. The only drawback to lime trees is that they rapidly outgrow their allotted space and bristle with long, needle-sharp thorns.

Limes take the place of the medium-sized, textured lemons used abroad. Lemons are indeed grown here but the large, warty-skinned, local varieties, though flavoursome, are scarce and very seasonal.

Grapefruits, sweet oranges (and sour oranges in particular), are just as useful in their own right and sometimes take the place of limes. The peel (leaves and flowers) of limes and all other Citrus contains fragrant, medicinal essential oils with limonene as the major constituent. Flavonoids, citral and a range of other active compounds are also present in lesser quantities. These substances provide the antiseptic, anti-inflammatory, anti-oxidant, decongestant and sedative properties that are so characteristic of this genus of plants. Citrus fruits have been the traditional, well endowed source of vitamin C, citric and other organic acids.

Limes

♦ SIMAROUBACEAE *Quassia family*

Quassia *(Picrasma excelsa)*
Other name Jamaican Bitterwood

Quassia tree

Quassia, in the form of small logs or wood chips, was a relatively well-known product, historically exported from Jamaica to Europe for use as a substitute for hops in brewing beer. Cold water or alcoholic extracts of the macerated wood and bark also provided simple, common, medicinal preparations or 'bitters'.

The trees, native to the Caribbean and Central America, can grow over 75 ft (25 m) tall in Jamaica. They have compound leaves and bear panicles of tiny, fragrant, yellow-green flowers that mature into small, dark coloured berry-like fruits. All parts of this species contain the intensely bitter quassin, other quassinoid triterpenes as well as various alkaloids. Jamaicans, regional tribespeople and others worldwide have employed the bitter infusions mainly as tonics, to stimulate the flow of digestive juices. Drinking water stored in cups made of quassia wood used to be one local method of getting a daily dose of bitters. Quassia solutions are anti-microbial, insecticidal (flies, mosquito larvae and lice) and are used to cleanse the bowels, reduce fever and kill intestinal worms and other parasites. Modern herbal use here is probably limited to the inclusion of quassia chips – called woodroot – in some 'roots' beverages.

The red-flowered South American shrub *Quassia amara* or Surinam Quassia, the original source of 'bitters', and the native *Picramnia antidesma* – Macary Bitter or Majoe Bitter – both close family members, can be used in much the same way as Jamaican Bitterwood. The name Quassia is derived from Quassi, an African healer who is said to have been the first user of the Surinam plant.

Quassia leaves and berries

♦ **MELIACEAE** *Mahogany family*

Neem *(Azadirachta indica)*

Neem tree

While hoary old Neem trees exist in Jamaica there is no accompanying history of local use of the species. This state of affairs will shortly change following the development of techniques for the mass production of

Neem plantlets (UWI – Biotechnology Unit) and plans for the delivery of saplings and education to our Jamaican farmers. Neem trees are a source of azadirachtin, a potent, natural pesticide that deters insects in many different ways. The chemical – which is non-toxic to humans, other mammals and birds – is produced in the seeds (most), the exceedingly bitter leaves and tree bark. Oil, easily extracted from Neem seeds, makes good lamp fuel; it is both antiseptic and insect repellent and is an ingredient for soap, toothpaste, shampoo and cosmetics. All plant parts and extracts also contain other bioactive triterpenoids (limonoids) as well as tannins and flavonoids. Some of these anti-microbial preparations are important in skincare and are applied to soothe and heal wounds, acne, dermatitis, chicken pox and other skin eruptions. Other infusions are medicinal, taken to treat fever, stomachache, malaria, parasitic worms and a host of other maladies. Neem twigs provide chew sticks – herbal toothbrushes.

Normal tree products are also harvested from Neem. They include timber, firewood (charcoal), fibre, resin, fertiliser (seed residues), mulch, green manure and fodder, but this last use is limited because of the bitterness of the foliage. Neems make great shade trees. The species originated in India where the simple, low-cost technologies outlined above have been successfully used for centuries. It is hoped that some of these major insecticidal, agricultural and herbal benefits may soon be realised in Jamaica and the wider Caribbean.

Neem flowers

Neem berries

This a fast-growing species, covered in a wealth of wind-stirred, tooth-edged leaflets and producing sprays of tiny, sweet smelling, white flowers from an early age. The resulting berry-like fruit ripen from green to yellow. Neem trees might occasionally be confused with the closely related **China Berry** or West Indian Lilac (*Melia azedarac*). These, however, are smaller, shrubbier plants also from tropical Asia, with similar fruit, but whose decorative, white flowers sprout unmistakable purple parts.

China Berry flowers

♦ EUPHORBIACEAE *Cassava and Spurge family*

Rosemary *(Croton linearis)*
Other names Jamaican Rosemary, Rock Rosemary

Seaside Rosemary bush backed by Seaside Grape plants

Jamaican Rosemary plants are branched twiggy shrubs that are widely distributed in the region. The aromatic foliage – dark, glossy-green above a pale, almost white matt finish on the lower surfaces – is somewhat akin in colouring to the leaves of their namesake, the Mediterranean Rosemary, *Rosmarinus officinalis.*

Seaside Rosemary shoots and fruits

There are two forms of *C. linearis* readily distinguished by leaf shape; those of the coastal variant are broad while the more cosmopolitan plants have relatively long, narrow leaves. Both types flourish on exposed rocky or sandy areas and can form extensive thickets up to 9 ft (3 m) tall. Rosemary flowers are tiny, bead-like blooms each with a crowd of white stamens. Some blossoms mature into small, lobed capsules.

Seaside Rosemary flowers – also showing felty-white underside of leaf

Breaking or bruising the young stems and leaves releases a strong, peppery fragrance that, when inhaled, can straightaway loosen the phlegm of head colds. Rosemary leaves are boiled as an important, scented ingredient for baths to wash patients, new mothers and to ease the symptoms of colds. Infusions are taken as daily beverages or drunk as teas to relieve colds, colic, fever, influenza and menstrual cramps. Other decoctions are applied to rheumatism, insect bites and used as hair wash. Dried twigs were once assembled into rustic brooms. Because of their lasting, penetrating aroma, Rosemary sprigs are used in magical and religious ceremonies. Specimens have been burnt in haunted dwellings to discourage unwanted ghosts (duppies) and, less controversially, employed as an insecticide principally against bed bugs.

Rosemary fruits

Rosemary shrubs – *Croton linearis* – must not be confused with the cultivated garden ornamentals commonly called 'Crotons'. These plants are *Codiaeum* species, a different genus from the same family.

Pepper Rod – *Croton humilis* – a small, obvious relation is a pungent, indigenous plant with unspecified herbal uses.

Pepper Rod flowering shoot

Castor Oil *(Ricinus communis)*
Other name Oil Nut

Castor Oil shrub

Ricinus communis is an easily recognised tropical species that was known to the ancient Greeks and Romans. Castor Oil plants are eye-catching and decorative, having smooth limbs and palmate foliage, but are short lived whether in tree or shrub form. These plants also exist in several varieties as instanced by the colours of stems, leaves and fruits that range from downy blue-green through pink to maroon. In Jamaica, the plants, which are part crop, part weed, are treated with some caution because of the known toxicity to humans, livestock and insects. All plant parts are poisonous to an extent but the main toxins, *e.g.* the lectin ricin, are concentrated in the bean-like seeds. This poison kills body cells by preventing the synthesis of proteins.

Red stemmed Castor Oil plant with male (yellow) and female (red) flowers

Stout, conical inflorescences bear the large, three-sectioned female flowers and spiny fruit capsules above the much smaller, yellow, male blooms. Oil Nut leaves, which appear as many-pointed, palmate fronds borne on long stalks, have an odd odour. They are used in bush baths and as poultices for the relief of head and stomachaches, arthritis, rheumatism and other ailments. Leaves are quailed (heated) to make them more pliable and then, with or without some oil, ointment or dressing, are wrapped around the afflicted body parts.

But, it is the big, fatty, mottled seeds that are most important. They are easily processed at home to extract a thick, almost black, malodorous liquid called Castor Oil. This raw, unrefined product does not contain the deadly ricin but carries the irritant substance named ricinoleic acid. It is this compound that makes Castor Oil into the ferocious laxative traditionally inflicted on school age children to purge them of intestinal worms. The oil is also used as an abortifacient. In other domestic applications the warmed oil is used to soothe ear and eye problems, clean and treat wounds, cure biliousness and constipation and condition skin and hair. Castor Oil has also served as good home fuel (lamps) and lubricant. As the seed pulp or mash remaining after oil extraction is rich in ricin, these residues must be disposed of rather carefully.

White stemmed Castor Oil plant with immature fruits

Oil Nut seeds contain a treasure trove of other phytochemicals most of which have beneficial activities. For example, there are several each that are cancer preventative, anti-tumour, anti-microbial, anti-inflammatory, anti-asthmatic, anti-cataract, anti-influenza and anti-dermatitic. Others, however, can cause cancers, allergies, convulsions, abortion and kidney damage.

Abroad, Castor Oil beans are the raw material for a wide range of industrial and pharmaceutical enterprises that make polymers (nylon), lubricants (*e.g.* Castrol), greases, soaps, varnishes, cosmetics, medicines and many other products.

Physic *Nut (Jatropha curcas)*
Other name Resurrection Tree

Physic Nut flower

These Tropical American plants are found in the wild here and are occasionally grown as hedging or windbreaks. The plants ordinarily reach 6–12 ft (2–4 m) in height and seem, at a glance, to resemble young cotton and other Malvaceae shrubs. However, *J. curcas* – and the entire genus *Jatropha* – is closely related to *Ricinus communis*, the Castor Oil shrub. Both plants possess similar biological and chemical properties. Physic Nut shrubs have smooth, thin, peeling bark over sap-filled stems, large trilobed leaves and bear clusters of inconspicuous, bell-shaped, yellow-green flowers. The female blooms give rise to bunches of fleshy fruits or capsules that, when mature, dry and split into three seed (or nut) bearing segments. All parts of the Physic (purging) Nut plant are toxic to some degree, but they are all used in various folk remedies.

Physic nuts or fruit

Fresh leaves – lightly crushed and quailed – are used to poultice bruises, cuts, ulcers and rheumatic sites. The sap (or latex), which is anti-bacterial and anti-fungal, is rubbed inside the mouth to treat abscesses, thrush and toothache. Sap and other preparations are also applied to skin eruptions such as eczema, ringworm and scabies. Plant decoctions are taken for bowel disorders – diarrhoea, dysentery and constipation – or used as diuretics and also imbibed to treat fever, jaundice, rheumatism and other ailments. The chewed seeds are strongly laxative and are employed to kill and expel intestinal parasites. Physic nuts are relatively large; nearly an inch (2.5 cm) long, very oily and the white seed flesh is pleasant to the taste. Unfortunately, the main plant poison, curcin (a lectin) and the purgative, oil based curcanoleic acid are both concentrated in the seeds. Although roasting is said to make the seeds less toxic, they remain potentially rather dangerous and should be treated with caution. Like those of the Castor Oil plant, Physic Nut seeds, if ingested in sufficient quantities, can cause severe gastrointestinal upsets, collapse and death. The common name Resurrection Tree refers to the false legend or stigma that finds the trees bleeding red if they are wounded during Good Friday. Of course, Physic Nut plant sap retains its usual reddish (or whitish) hue all year round.

Jatropha curcas plants now exist worldwide and are used medicinally in other Caribbean countries and extensively in Africa. The yellow oil extracted from the seeds is used commercially in fuels, candles, medicine and soaps.

Flowers of the Belly-Ache shrub

Belly-Ache Bush, Cassada Marble or Wild Cassada are common names for *Jatropha gossypiifolia*, a much smaller, hairy and equally toxic plant. The leaves are deeply lobed and the young growth – particularly that of the red-brown variants – is a rich, glossy chestnut colour. These plants, about 3 ft (1 m) tall, often grow together in crowded patches on rocky, gravelly or sandy areas close to the sea. Their small, attractive blooms have yellow centres, dark red petals and develop into lobed fruit each with three, tiny seeds. The leaves are commonly boiled down to make traditional, all-purpose herbal medicines. Otherwise, Belly-Ache Bush – also known as Wild Physic Nut in parts of the eastern Caribbean – is used in much the same way as Physic Nut plants.

An endemic species, Wild Oil Nut *(J. divaricata)* is frequent in thickets on limestone. Several other species of *Jatropha* – Bottle Plant, Coral Plant, Gouty Foot (*J. podagrica*), Coral Plant (*J. integerrima*) and French or Spanish Physic Nut (*J. multifida*) – are cultivated here as ornamentals.

Physic Nut (left) and Castor Oil (right) fruit and seeds

Seed-pon-back *(Phyllanthus amarus)*
Other names Carry-me-Seed, Egg Woman

These small, widely-distributed, tropical weeds will spring up anywhere, given a good supply of moisture and disturbed ground. There are several very similar species of *Phyllanthus* that share the same looks, local names and a few uses. The lime green plants are usually less than 18 ins (50 cm) tall and bear tiny, male and female flowers as well as the signature, egg-like fruit on the undersides of the closely matched, alternate leaves.

Leaf teas are taken to stimulate the appetite, cure stomachache and treat mouth problems. Substances identified from Phyllanthus species include methyl salicylate (wintergreen), terpenes (limonene) and bitter alkaloids (phyllantine). Plant extracts in the main are anti-bacterial, anti-viral, anti-fever

Seed-pon-back plant

and anti-inflammatory. The weeds are known elsewhere in tropical South America as shatterstone and the herbal preparations employed as antidotes to disorders of the liver, gallbladder, kidney, bladder, digestive and urinary tracts.

In Jamaica, the introduced cultivated tree, *Phyllanthus acidus*, bears those small, yellow, tart fruit: childhood's delight called Jimbelins.

Seed-pon-back fruits on the undersides of the branches

♦ RHAMNACEAE *Buckthorn and Coolie Plum family*

Chew Stick *(Gouania lupuloides)*

Chew Stick vine climbing on a tree

Chew Stick comes home from the forest coiled like rope, bundled over the head and under one shoulder. This ever popular, woody, climbing vine is found in tropical America and the West Indies. Here in Jamaica, though somewhat thinned by over collection, the vines are still found in thickets and woods, extending up to 36 ft (12 m) on host trees.

Chew Stick branch showing characteristic tendrils

All parts of this bitter-flavoured plant go to help create lotions, gargles and washes for use on sores, ulcers or infections of the gums, mouth and throat. Some infusions are taken as aphrodisiacs or to treat stomach complaints. However, the most enduring function of this vine (wis) is as chew or chaw sticks – an African heritage. The stems are chopped into handy sized lengths, soaked, and then one end of each stick is beaten, chewed or frayed into bristles. This rough brush is used to clean the teeth and massage the gums with help from the resulting bitter, anti-bacterial, astringent and frothy in-mouth solution.

The dried vine, once exported to Europe and America for inclusion in tooth powders, is now used here in the 'Chew Dent' series of stern local oral hygiene products. Plant extracts, flavoured with peppermint and wintergreen, are used in the manufacture of a green, gel toothpaste and a similarly coloured mouthwash.

Chew Stick's foamy bitter principles are due to saponins that were once used to flavour homemade drinks and contribute to the fermentation of ginger beer and other brews.

◆ TURNERACEAE
Turnera family

Ramgoat Dashalong
(Turnera ulmifolia)
Other names Ramgoat
Nation, Ramgoat
National, Buttercup

This is one of many relatively short, bushy *Turnera* species that are indigenous to the American tropics. The plant is a weed here, found in waste places that range from forests to fields, pastures, rocky and sandy coastal areas. The branching shrubs

Ramgoat Dashalong leaves and flowers

are generally 3–4 ft (90–120 cm) tall and carry alternate, saw-toothed, lance-shaped leaves that are crowded on the tips of young twigs. This foliage has a rough texture but it is attractive and is usually decorated by a series of beaming, bright yellow, flat-petalled blooms. Ramgoat Dashalong is one of the few common 'bush' plants that have been willingly planted in up-market gardens. The shrubs are inclined to become leggy but they are very hardy and highly decorative when in full flower. The blooms maybe over 2 ins (5 cm) wide and are renewed daily. Another reason for this interest is that the plants are regarded as good tea bushes. Crushed leaves have a distinct aroma – thymol and other terpenoids – and the young growth is frequently brewed up to make a golden-green solution that is taken for many purposes. Local folk have these teas as general beverages, or take them to fight colds, depression, digestive and other upsets – some clearly hope that the drinks will act as aphrodisiacs. In Brazil, plant extracts have long been used to treat ulceration and related diseases of the gut. Some scientific research suggests that these infusions contain certain flavonoids, compounds that prevent or alleviate ulcers. *Turnera ulmifolia* is also known as West Indian Holly, Yellow Alder and Sage Rose around the region but the connection with frisky, Jamaican male goats remains speculative.

Jamaica is home to three other similar *Turnera* species – the rare *T. pumilea*, the endemic *T. zeasperma* and the smaller *T. diffusa*. This last is synonymous with *T. aphrodisiaca*, the famous 'Damiana' – a traditional, high profile herb plant abroad that is said to be reliably anti-depressant, anti-diabetic, calming and a tonic but, also a stimulant and aphrodisiac to both sexes, amongst its many other virtues.

◆ **BIXACEAE**

Annatto *(Bixa orellana)*
Other names Natta or Natto

Annatto is a quick growing dye and spice plant that was, in the past, a major, international source of red and yellow pigments for use in foodstuffs, cosmetics and textiles. The port and town of Annotto Bay on the northeast coast of Jamaica in St Mary took its name from the Annatto industry, which once flourished in the area. The

Annatto flowers

shapely, medium-sized shrubs or trees bear rose-like, pink or white flowers that mature respectively into bristly, triangular fruit pods that are either red-brown or green in colour. It was the invention abroad of cheap, artificial colouring that eventually squashed the export business and reduced Annatto plantations back to the original cottage gardens and those shrubs that grow wild in hedges and on waste ground.

Annatto Yellow fruit

Annatto Red fruit

All parts of this plant are used elsewhere in the Caribbean and tropical America to make a wide variety of medicinal preparations, but in Jamaica, it is the seeds that are most important. Inside each Annatto fruit is a generous provision of vitamin rich, orange-red pulp and seeds. The latter are still used – dried – to colour and flavour cooking oils, soups, stews, rice and other dishes. Elsewhere, seed extracts are made into antidotes for cassava poisoning and the large, soft, oddly-scented leaves are put in baths and made into teas to treat worm infestations, inflammation, colic or heat. Active plant chemicals such as Bixin and other carotenoids are astringent substances, effective at healing wounds and burns.

Bixa orellana is the sole species of the single genus and family, restricted to South America and the West Indies. It provided the red body paints so favoured by ancient (and modern) Amerindians for skin protection and ritual decoration. The plants, however, are prone to mildew and often host stinging ants that come to raid the large, stem nectaries.

The fortunes of Jamaican Annatto farmers may yet rise again in line with some recent, helpful developments. These include the growing, worldwide demand for more natural, plant-based food additives and colourants, the arrival of a local, Annatto-derived, herbal skin product (Cumsee Ointment) for treating stubborn, chronic leg sores (ulcers), and the holding of the first ever Annatto festival in its home town of Annotto Bay.

◆ **CUCURBITACEAE** *Cucumber and Squash family*

Cerasee *(Momordica charantia)*

Cerasee leaves and flowers

Cerasee is easily Jamaica's premier herbal plant in both availability and popular use – only topped by the prohibited marijuana (ganja). The lovely-leaved weed, universal at all but the highest altitudes, is regarded as both a tonic and general cure-all. Cerasee plants make prolific, thin-stemmed vines that rampage along the ground, climb fences and thread through other plants with the help of coiling tendrils. Most Jamaicans have experienced at least one mug (or dose) of sweetened tea made from the young leaves and stems of this plant. The relentlessly bitter brew can be strongly purgative and is credited – among other miracles – with curing bad blood, bad skin, fevers, diabetes, colds, gripes and all ills associated with the belly. Folk wisdom dictates that the seeds should not be used and that a course of treatment should not last longer than 7–9 days.

Fresh leaves of Cerasee and Quaco Bush (*Mikania micrantha*) were traditionally pulped together in hot or cold water to produce an acrid, dark green, sudsy mixture. This extract was added to baths to help maintain smooth skin or treat the distressing symptoms of acne, eczema, pimples, rashes and other external complaints. Formerly, when detergent cleaners were scarce, the herbal pair were often deployed to scrub wooden furniture, floors and ink stained school desks.

Unripe Cerasee fruits: the two large ones are from different cultivated vines, the small ones are from a wild vine

Ripe Cerasee fruit on a wild vine

Ordinary Cerasee vines produce a succession of small, bright yellow blossoms. Female flowers mature into petite, warty-skinned fruit. These become very sweet, edible and orange in colour when ripe and, soon after, they rupture wide apart in a three way split. Birds – and children – feast on the slimy, red aril that covers the exposed seeds. The fruit of a much larger, cultivar (Caraaili) may be 6–12 ins (15–30 cm) long, and are cooked and eaten as a bitter green vegetable. Cultivated and wild Cerasee vines will interbreed seamlessly to produce fruit in a range of intermediate sizes.

Large (cultivated) and small (wild) Cerasee seeds - some covered with red aril

Local research has confirmed that in addition to the predictable glut of vitamins, minerals, amino acids and aromatic oils, the plants are also rich in therapeutic agents. Fruit extracts, in particular, contain insulin-like compounds, HIV-inhibiting proteins, laxative saponins, bitter alkaloids and other substances that are anti-bacterial, anti-parasitic and abortifacient. Therefore, Cerasee juice and infusions can, for example, reduce blood sugar and act against biliousness, cancer, HIV virus (in the test tube), leukaemia, psoriasis and mites. Persons who should not take Cerasee include young children and those who are hypoglycaemic, *i.e.* suffer from low blood sugar.

Cerasee is traditionally much used in the Caribbean, Africa and Asia. Today, the herb and its herbal products are widely traded around the world.

◆ **MYRTACEAE** *Myrtle and Guava family*

Pimento *(Pimenta dioica)*
Other names Allspice, Clove Pepper, Jamaican Pepper

Pimento bark

Pimento flowers

Pimento is a tree herb with plants that can top 45 ft (15 m) in height. They are native to Jamaica, the Caribbean, Central America, and now introduced to many other tropical places. The trees, like other family members, have smooth brown bark, which peels off in long pieces, giving the trunks that distinctive, attractive, stripped look.

Pimento berries

Pimento – like Ginger – has been economically important to Jamaica since the beginning of the sixteenth century. There is a large, profitable trade in the dried, unripe berries and essential oils extracted from both the fruit and leaves. These and all other plant parts are wondrously scented with what seems like a mixture of cinnamon, cloves, juniper, nutmeg and pepper giving rise to the inclusive name of Allspice! All Pimento trees can bear flowers and fruit. They are not dioecious – separate male and female plants – as was thought when the species was first named.

Pimento products are used here and worldwide as a spice and a preservative in both savoury and sweet cooking. Pimento is the obligatory ingredient in 'jerk' seasoning. Ripe berries, which are available from August to September – flavoured with cinnamon, cloves and ginger – go to make the esteemed bright red, alcoholic drink or cordial called Pimento Dram.

Herbal therapies centre on using pimento as a warming agent to stimulate the digestive, circulatory, nervous and respiratory systems. Pimento teas and decoctions made from the leaves and young twigs are taken to improve the appetite and treat digestive upsets such as flatulence, indigestion, nausea and vomiting. The leaves are put into baths, and pounded green berries and leaves are used as a mash for plasters and poultices. These are applied to the joints and body parts suffering arthritis, muscle cramps, rheumatism and other painful conditions. Pimento extracts are also taken to relieve coughs, colds, influenza, bronchitis, pneumonia, nervous depression, tension and stress.

The main chemical constituents of pimento include eugenol (oil of cloves) and cinnamaldehyde. These substances and others are strongly analgesic, anaesthetic, anti-fever, anti-inflammatory, CNS-stimulatory, carminative and sedative among other pharmacologically beneficial properties; all of which support the traditional use of Pimento in herbal remedies.

Pimenta racemosa is the Bay Rum Tree whose leaves, steeped in white rum, provide that stinging, fragrant, volatile tincture which is rubbed on at home for minor aches and pains, cuts, bruises, chills, fevers and other discomforts.

There are also two endemic *Pimenta* species, each with its own aroma, neither of which is like Allspice. One is a subspecies of *P. dioica* whose members mostly possess a citrus-like scent. The other, *P. jamaicensis*, is known as Wild Pimento.

Guava *(Psidium guajava)*

Guava branches showing the smooth bark

Guava plants bear the fragrant, pink-fleshed fruits with gritty seeds that are the sweet delight of many a bird, bat, rat and child and the longtime source of jams, jellies and juice. They are indigenous shrubs or small trees that are sometimes cultivated but which grow naturally in abundance on old pastures and wasteland. These plants are regarded as invasive weeds in parts of Florida. Most well-grown Guava shrubs exhibit the peeling bark and smooth, red-brown patterned stems characteristic of other family members. Flowers are attractive starry white blooms with five petals and a mass of stamens.

Guava flowers, leaves and unripe fruit

Teas made with guava leaves (and buds) are usually taken for gut problems: stomachache, nausea, diarrhoea, dysentery, vomiting and are given to children to expel worms. Infusions, made by steeping the leaves in hot water, make anti-microbial solutions that are used for treating skin troubles such as abscesses, boils, cuts and ulcerated wounds.

Guava leaves contain many pharmaceutically active compounds like flavonoids (quercetin), essential oils with eugenol and other terpenoids and high concentrations of tannins. These substances provide the anti-bacterial, anti-fungal and anti-oxidant activities that are effective against the domestic infections listed above.

The following three wild guava species bear smaller but edible fruit. They are *P. guineense*, the endemic mango-flavoured Mountain Guava *P. montanum*, and the Purple or Strawberry Guava *P. cattleianum*.

Bunch of ripening guavas

Ripe guavas split to show pink flesh and seeds

◆ APIACEAE *Carrot family*

Fit Weed *(Eryngium foetidum)*
Other names Spirit Weed, Culantro, Stinking Eryngo

Rosettes of Fitweed leaves

Both scientific names describe Fit Weed. *Eryngium* is the genus of Sea Hollies – plants resembling thistles – while *foetidum* is the Latin for an unpleasant smell or stink. In this case, unusual, might seem more accurate.

Fit Weed is used as instant herbal smelling salts. It comes in handy when a sharp, reviving scent is needed. The small, prickly herb grows in damp, shady places and favours the edges of paths and places adjacent to old concrete. The plants are normally biennial and produce just a rosette of long, toothed leaves at first. When crushed, these leaves emit a strong smell reminiscent of coriander (cilantro) and are vigorously rubbed on to persons (usually children) suffering from fits, fainting spells or convulsions. Later, the central stem elongates up to 1 ft (30 cm) in height and the plants develop tiny, cone-like inflorescences at the ends of short branches. Each flower head is seated in a ring of four to five, long, pointed sepals.

Teas made from the leaves and roots are used primarily to combat seizures, fight colds, fevers and stomach upsets. Other conditions treated with the herb include asthma, animal bites and stings, depression, high blood pressure, pneumonia, rheumatism and intestinal worms. Fitweed leaves are reportedly rich in minerals (calcium, phosphorus and iron), vitamins A, B and C and a

Fitweed plants with branched flowering stems

Fitweed flowering heads

potent volatile oil. Comparable essential oil extracted from coriander leaves contains mainly linalool, coumarins and triterpenes.

Eryngium foetidum is used extensively in other West Indian islands, Central America and Asia as a pot or seasoning herb. The plant is widely known as Culantro – not to be confused with the related, parsley leaved cilantro – and there are over 60 other commonly used names, for instance Shado Beni in Trinidad and Recao in Puerto Rico. Other versions popular in the Eastern Caribbean include Chadron Benee, Shadon Beni, Shadow Benny and even Black Benny. These are all corruptions of the French 'Chardon Beni' or blessed thistle. The pungent Fit Weed leaves are the bases of chutneys and sauces and they flavour salads, vegetable and meat dishes. There is a brisk trade in the fresh, green leaves from Trinidad & Tobago to ethnic populations in big cities in the UK and North America.

Eryngiums are native to both Europe and South America. The odourless, evergreen, European relation, the Sea Holly – *E. maritimum* – grows wild on sandy, coastal areas. Their cooked roots were something of an aphrodisiac for ancient Britons. Today, the young, green parts are eaten as a vegetable and the roots candied as well as being made into a tonic for urinary problems.

♦ **SAPOTACEAE** *Sapodilla and Star Apple family*

Naseberry *(Manilkara zapota)*
Other name Sapodilla

Naseberry tree

Naseberry flower

Naseberries

Naseberry plants, which originated in Central and South America, have long spread to the Caribbean, Florida and the rest of the tropical world. There are several different types and cultivars differentiated mainly by the shape, size, texture and sugar content of the fruits. The plants here are seen most often as shapely, medium-sized, latex-filled trees that are covered in glossy foliage and obscure bell-shaped white flowers. These, and the resulting abundant crops of scurfy, brown skinned fruit, are produced throughout the year. Ripe fruits have fragrant, golden-brown, honeyed flesh and are regarded as natural treats by man, beasts and birds alike. Naseberries are eaten fresh on the go as snacks, or as desserts at meals and have been included in drinks, ice creams, jams, jellies and other preserves. Their black seeds, which contain small quantities of the triterpenoids saponin and the bitter sapotinin, should not be ingested as these compounds can cause stomach upsets and vomiting. Although these plants flourish in the wild – along with other fruit trees such as breadfruit, mango, sweet and soursops – they are also all cultivated in gardens, smallholdings and on farms. Naseberry trees, particularly those grown on plantations in Mexico, India, and the Far East primarily provide sweet fruit, but the plants have secondary roles as herbal resources.

In Jamaica, teas made from the leaves are taken as nerve tonics and used to treat colds, coughs, fevers and influenza. Naseberry bark and young fruit are rich in tannins and decoctions made from these plant parts are employed to control fever and limit the debilitating effects of diarrhoea, dysentery and other upsets of the digestive system.

In their native lands, naseberry trees are important sources of the sticky, white sap called 'chicle' which was, until recently, the main ingredient of chewing gum. Chicle, which contains 15 per cent rubber and 38 per cent resin was historically dried and chewed by local people and was used as a folk filling for cavities in decayed teeth. The hard, red heartwood of the Naseberry trees is a valued timber product.

◆ APOCYNACEAE *Dogbane and Oleander family*

Ramgoat Roses *(Catharanthus roseus)*
Other names Periwinkle, Vinca, Old Maids, Bright Eyes

Ramgoat Roses are natives of Madagascar, but are now distributed far and wide to other tropical places. They are adaptive, shrubby, scrambling weeds that can endure harsh conditions in gardens and the waste places they so frequently inhabit. Well-nurtured plants have shiny, dark leaves and carry the familiar flat, five-petalled flowers in shades of red, pink and white with more or less matching or contrasting 'eyes'. These simple but attractive blooms are produced nearly all year round.

Periwinkles contain the irritant white sap common to family members, and the plants are neither vase nor goat friendly. In the first instance, flower stems develop rank odours when left in water and even our seasoned troopers (goats) give the weeds a miss while foraging for food.

The green shoots have long been used to make traditional herbal remedies. Leaves and stems were boiled down to make bitter medicines mainly for diabetes and high blood pressure. Other ailments treated included arthritis, diarrhoea, fever, and nervous conditions. Preparations were also used to stop bleeding from the mouth and nose as well as excessive menstrual flow. An infusion of the flowers was said to soothe coughs and sore throats and chewing fresh leaves was expected to ease toothaches.

Nowadays, there are warnings on the use of these herbal infusions. *Catharanthus roseus* plants contain over 90 alkaloid chemicals (monoterpene indole alkaloids mainly) and some of them – catharanthine and vindoline from root extracts – are indeed anti-diabetic (hypoglycaemic) compounds. Vinblastine and vincristine (dimeric alkaloids) have been isolated from the plants and made into drugs to treat childhood leukaemias, melanomas and various other cancers that are found in the breast, lung and uterus. However, these last two phytochemicals are also very poisonous substances that can cause gastrointestinal and nerve damage and may mask the true sugar content of urine; hence the ban on the use of herb teas brewed from Ramgoat Roses.

Ramgoat Roses plant growing
on a beach

◆ BORAGINACEAE *Borage and Clammy Cherry family*

John Charles *(Cordia globosa)*
Other names Gout Tea, Black Sage, Wild Sage

John Charles is plentiful in open, dry areas like some old pastures and churchyards of hilly Manchester. It is normally a rather short, very slow growing, *Lantana*-like shrub with rough surfaces and dark purplish stems that stand up to 3 ft (1 m) tall. However, these plants can scramble with support up to 18 ft (6 m) in height. Although *Cordia globosa* is also found throughout the rest of the West Indies, Florida, northern South America and Central America, it is listed as endangered. In some quarters the plant is called Butterfly Sage in recognition of its appearance and usefulness as a nectar provider.

This John Charles shares its name with at least two other herbal plants – *Lantana urticifolia* and *Hyptis verticillata* – but its toothed, scabrid leaves are not aromatic, unlike those of both namesakes. The foliage and young stems have raised areas with flattened hairs that make the plants harsh to the touch. Nevertheless, John Charles is well known in Jamaican herb lore as a good tea bush. Infusions made from the leaves are mainly taken for chest colds, marasmus, mucous congestion, tightness of the chest and gout; sometimes in combination with Jackney and Search-mi-Heart. The teas are also used to ease menstrual cramps, control internal and external bleeding and are added to baths as treatment for skin conditions.

John Charles flowers and leaves

John Charles plant in fruit

Flaring, milky-white, funnel-shaped flowers are borne in terminal, globular clusters and some of these beautiful blossoms develop into attractive oblong, red berries. Birds eagerly eat these succulent treats. *Cordia globosa* is just one of half a dozen or so similar, local, shrubby *Cordia* plants.

Cordia bullata in fruit

Scorpion Weed *(Heliotropium indicum)*
Other name Wild Clary

These are common, tropical weeds that grow to a maximum height of about 3 ft (1 m). The plants have closely alternate, toothed leaves and they bear distinctive flowering spikes that arch and curl somewhat like the tails of scorpions. These inflorescences have flattened tips that carry two rows of exquisite, tiny, yellow-eyed flowers whose corollas are either white or lavender coloured. Scorpion Weed, the very similar Dog's Tail (*H. angiospermum*) and the smaller *H. curassavicum*, grow wild in cultivated and waste places. The plants pop up in fields, gardens, pastures, sandy and rocky coastal areas, and roadsides where they flower and fruit all year long. Fertile seeds – nutlets – from the rounded fruits germinate easily to repeat the life cycle.

Scorpion Weed plants flowering and fruiting

Scorpion Weed flower showing its coiled shape

Herbal Plants of Jamaica

Scorpion Weed – lavender-flowered species

Scorpion Weed – white-flowered species

Leaf juice or plant decoctions are made into lotions for bites, boils, stings, skin ulcers, wounds and eye ailments. The preparations are applied externally: they should not be swallowed because of the poisonous nature of the Scorpion Weed plant. All *Heliotropium* species – and most members of the Boraginaceae family – contain certain toxic compounds called pyrrolizidine alkaloids. These substances can cause cancer, liver damage and eventually death to humans and grazing animals if taken in large enough doses or over a long period of time. Nevertheless, Scorpion Weed has been used as a medicinal herb in India, China and Africa for centuries.

The name Wild Clary (Clear Eye or Eyebright) was originally applied in Europe to species of *Salvia* whose leaf extracts and mucilaginous seeds are still used as eye drops to counter inflammation and wash the eyes clear of irritant particles.

◆ **VERBENACEAE** *Verbena and Lantana family*

Colic Mint *(Lippia alba)*
Other names Colon Mint, Guinea Mint, Cassava Flower

Colic Mint shrub

A vigorous, straggling, rather untidy shrub, *L. alba* has many pale, round stems that may stretch 6 ft (2 m) tall. Colic Mint – heir to many local names – although grown in some gardens, largely exists as wild plants in the scrub vegetation bordering our southern coasts. The species is common to the West Indies, Central and South America, in particular Argentina and Brazil where the plants are important, cultivated medicinal herbs.

Colic Mint flowers

At every stem node the plants bear clusters of small, sweetly peppermint scented, hairy leaves and a pair of fleshy, stalked flower spikes. Each resembles a pointed, minature pineapple less than 1 in (2 cm) long and consists of a head of compacted bracts bearing small tubular flowers. The corollas are white (sometimes pink or purplish) and they surround a yellow centre. The entire plant is incredibly aromatic and releases a wonderful fragrance at the slightest touch.

Colic Mint fruit

The leaves here are used to brew minty teas to soothe colicky infants, check flatulence, 'cut' fever, treat insomnia and cure venereal disease. Elsewhere, the infusions act as anti-depressants, digestives and anti-spasmodics and are also taken as remedies for colds, fever and influenza.

Colic Mint plant extracts contain flavonoids, alkaloids and important essential oils that include cineole, limonene, myrcene, geranial and neral: substances that are analgesic, anti-microbial, bronchorelaxant, fragrant, expectorant and sedative.

Vervain *(Stachytarpheta jamaicensis)*
Other names Vervine, Porterweed, Porterbush

Vervain flowers

Vervain shoot

Jamaican Vervain is a low growing, rambling weed that is altogether very common in open, waste or disturbed places. The plants seem not inclined to form thickets and are usually found singly or in small groups. They are native to Tropical America and the West Indies but have now spread to the Pacific Islands. Vervain leaves are interesting, being mainly opposite, oval, toothed and borne on square, green or purplish stems. But it is the fleshy flower spikes – borne year round – that are distinctive. These may extend up to 20 ins (55 cm) in length and bear successive groups of 4–5 small, tubular, bluish-purple flowers – one whorl at a time – along the spike. This leaves the greater length of each spike bare of flowers at all times and is the origin of the common name 'Rat Tail,' used for *S. jamaicensis* elsewhere. The tongue-tripping generic name – derived from the Greek – is also based on the flowering stalks. *Stachys* is a spike and *tarphys* is 'thick' which describes the crowded arrangement of the flower buds on the spike. Porterweed (or Porterbush), on the other hand, refers to a traditional use of the plants elsewhere to brew a foamy, porter-like beverage. (Porter is an old-fashioned London beer).

The leaves and roots of Vervain are used in many ways to treat an extensive array of discomforts and ailments. They may be crushed and used topically to soothe skin irritations. Plant parts may also be juiced or boiled often in mixtures with other herbs. Plant and leaf extracts in particular contain caffeic acid, flavonoids, saponins, tannins and other beneficial and protective phytochemicals. These substances are collectively analgesic, anti-inflammatory, anti-oxidant, anti-viral, bactericidal, immunostimulant and vasodilatory; activities that would seem to support folk use of Vervain in baths, as a tonic and as treatment for blood conditions, colds,

Vervain flowers on purple spike

fevers, headaches and sores. Plant extracts are also used to treat asthma, diabetes, eye-problems, high blood pressure, jaundice, nervous complaints, sleeplessness and, in conjunction with soursop, wormseed and castor oil used to expel worms. Some preparations are employed as abortifacients, emetics and purgatives. Packets of the dried, shredded herb are now available at various commercial outlets. Other than being said to be bad for sheep, Vervain is not known for causing toxic side effects.

The several other local species of the genus are similar to Vervain in many ways and probably interchangeable in Jamaican herb lore. *Stachytarpheta adulterina* and *S. angustifolia* are both blue flowered but *S. cayennensis* has tiny, near white, starry flowers while *S. mutabilis* sports bright, rosy pink blooms.

Vervain – red flowers

♦ LAMIACEAE (LABIATAE) *Mint and Joseph's Coat family*

French Thyme *(Plectranthus amboinicus syn. Coleus amboinicus)* Other name **Soup Mint**

French Thyme is usually confined to borders, pots or other containers around the garden. If left to themselves, the plants can quickly proliferate to a spreading sprawl of succulent light green herb. Their fleshy, hairy leaves are quite triangular with pronounced saw-toothed edges. When bruised or rubbed between the hands these leaves almost liquefy as they produce a strong, refreshing fragrance midway between mint and thyme. These crushed leaves are used to combat tiredness and treat faints and fits. Leaf teas are taken in response to fever, headaches, 'nerves', heart and urinary problems. Shredded leaves may be added to soups and stews but French thyme is not heavily used in cooking here.

The plants bloom en mass for an extended period during the first half of the year. Elongated flower spikes produce delicately angled, lilac blooms that are extraordinarily similar to the bluish-purple inflorescences of the closely related *C. scutellariodes* (syn. *C. blumei*), Coleus or Joseph's Coat.

Plectranthus amboinicus has spread worldwide from its origins in the Moluccas Islands (Indonesia) and is known by a variety of other names such as Country Borage and Cuban Oregano. The herb is cultivated abroad for its essential oil, of which the most important component is thymol at 94 per cent. Plant extracts therefore have anti-microbial, anthelmintic, anti-spasmodic and expectorant properties.

French Thyme shoot and flowering spike

French Thyme flowers

Piaba and related herbs *(Hyptis* spp.*, Leonotis* sp.*)*

Hyptis species, common in tropical and subtropical countries, are wild shrubs that have paired, saw-edged leaves, ridged stems and unusual flowerheads. Some are invasive weeds (Hawaii) while others are prized medicinal plants (Africa). The name Piaba stirs a chord with many Jamaicans who have used the plants and know the popular verses that were sung about old time herb vendors, 'She had Man Piaba, Woman Piaba, Tom Tom Fall Back and Lemon Grass' and so on.

Woman Piaba plant

Woman Piaba shoot

Woman Piaba (*Hyptis pectinata*) plants grow 3–9 ft (1–3 m) tall and flourish from sea level up to mid-altitudes. Their clustered, woolly inflorescences are made up of prominent calyces and tiny purplish blooms; the whole exuding a very pleasant aroma when stroked. The felty, heart-shaped leaves are brewed into teas to ease childbirth pains, a distinctively female application that is also used in parts of Africa. Leaf teas are taken for the usual round of complaints, head and stomachaches, fever and chest colds. Leaf juice is applied to sore throats, cuts and ulcers. There is some concern that chemical substances present in these extracts may slow the heart rate.

John Charles foliage

John Charles flower spikes

John Charles flowers

John Charles or His Hog Money (*H. verticillata*) is the most abundant of the *Hyptis* herbs. These plants grow up to 9 ft (3 m) tall and bear long, arching, white-flowered branches. The slender, toothed leaves – much used in bush baths and teas – contain several chemicals, in particular podophyllotoxin, an anti-viral, anti-tumour compound and rosmarinic acid which is effective against gastrointestinal disorders and skin infections. The aromatic **Picknut**, Pignut, Spikenard or Wild Basil (*H. suaveolens*) has broad leaves and short, purplish flowering heads. **Ironwort** or Wild Caesar Obeah (*H. capitata*) grows thickly in damp places. The shrubs have hollow stem sections, large lobed leaves and small, stalked, solitary, white flowerheads. The anti-microbial and liver protective plant extracts are reportedly used in Ecuador to treat fungal infections and in Taiwan against asthma, colds and fever. But, despite the suggestive Jamaican names, I have not discovered an authentic local herbal use.

Extracts of *Hyptis* herbs contain rafts of medicinally active compounds including essential oils with carophyllene, camphor, menthol, pinene, thymol and others. These substances are analgesic, anti-oedemic, anti-bronchitic, anti-inflammatory, anti-microbial, anti-fever, anti-tussive, decongestant and expectorant and are used in various herbal preparations to treat cancer, colds, fever, digestive problems, skin diseases, marasmus and other maladies.

Man Piaba, though not strictly a *Hyptis*, is named – in Barbados – as the closely related *Leonotis nepetifolia*. This is a scarcely branched shrub up to 6 ft (2 m) tall with similar, *Hyptis*-like, ridged stems but bearing large, stacked, globular flower heads. They are orange flowered and four to six times the width of the small *H. pectinata* equivalents. Also called Bald Bush and Christmas Candle Sticks here, these flowering stalks are particularly attractive live or dried. Teas made from the scented leaves and roots are valued as male libido-enhancing tonics.

Ironwort plant

Man Piaba
flowering shoot

125

Black Mint *(Mentha x piperita)*
Other names Peppermint, Broad Leaf Peppermint

Black mint plant   Black mint flowers

Black Mint is a standard, back door, garden herb that is usually restrained in containers. It is a sterile, triple hybrid, European cultivar that can be propagated only, but easily, by cuttings. Creeping, rooting stems bud off erect shoots that may top 2 ft (66 cm) in height. *Mentha* plants have square stems and in this particular herb they are an attractive, reddish-purple colour. Black Mint sports bright green foliage consisting of alternate, scalloped, heart-shaped leaves that, when bruised, smell very strongly and pleasantly of peppermint. The plants can flower freely in summer producing tightly-packed clusters of tiny, pinkish flowers. Crushed leaves are rubbed about the head and body to refresh and revive flagging spirits, soothe headaches and also to help clear blocked respiratory passages. The chief use of this herb, however, is in the making of leaf teas that are taken to treat a range of digestive disorders. These include nausea, stomach gas, irritable intestines and troublesome gall bladders.

The active substance in Black Mint is mainly an essential oil, the very widely known peppermint oil, whose chief components are menthol, its derivatives and other monoterpene compounds. The remaining constituents include flavonoids, organic acids, resins and tannins. This oil has indeed the anti-microbial, anti-spasmolytic, carminative, cooling, choleretic and pain relieving properties that are so important in the many herbal preparations.

Barsley *(Ocimum campechianum* syn. *Ocimum micranthum)*

Flowering Barsley plants

Barsley or Baazli, Jamaica's all-island, wild basil, is a low growing, perennial herb with square stems and distinctive flowering heads. The plants do spread sideways with a maximum height of around 20 ins (50 cm). When bruised, the soft, paired, light green leaves reward with a sharp, warm aroma reminiscent of basil, mint and citrus. Local folk drink Barsley tea and beverages to aid digestion and combat fever and pains, but the plant is not much used as a salad herb. The pungent crushed leaves are employed as body rubs and, in addition, all green parts are thrown in to scent bush baths. Like other highly aromatic plants, Barsley is associated with rituals and ceremonies such as burials and the defence against Duppies, our ethnic ghosts.

Barsley flower spikes

Barsley mature flower spikes

The plants produce short-lived, pale blue or white, tubular flowers all year round. Their tiny, purple-splotched blooms seem constrained by the persistent, leafy calyces. The upper surface of each mature, tube-like calyx is hollowed out like a miniature bathtub, while the lower edge develops a stiff fringe. These erect, bushy inflorescences become even more distinctive in maturity, persisting as attractive, characteristic, wood-brown spires. As Barsley plants can quickly become leggy with declining foliage and flowers, they should be rejuvenated periodically by taking cuttings or by planting the numerous ripe nutlets. Although this wild basil goes by two other common names – Nunu and Tulsi, which are derived from Africa and India respectively – the plant is native to the American tropics and subtropics. The evocative scents, flavours and medicinal functions of Barsley – and all other *Ocimum* species – are due largely to the chemical compositions of their inherent essential oils. These volatile oils contain specific mixtures of highly active terpenoid compounds such as cineole, linalool, eugenol, carophyllene, pinene and thymol.

African Tea Bush

African Tea Bush shoot

African Tea Bush flowers

African Tea Bush or Balsam (*Ocimum gratissimum*), a close but generally rare relation is abundant by the roadsides in at least one local river valley. Here, the plants are 3–5 ft (1–1.5 m) tall but they can extend up to 9 ft (3 m) abroad, prompting the common name of Tree Basil. The plants are from Tropical Asia but are widely grown in parts of Africa and sparsely in some other Caribbean islands.It is a much taller shrub than Barsely with correspondingly large, saw-edged leaves and long, lax, flowering spikes. The flowers, which are incredibly tiny, seem to have white corollas but it is their long, protruding, yellow tipped stamens that are obvious. The foliage is also strongly scented but it has a heavier, more musky aroma than that of Barsley. African Tea Bush leaves, like those of Barsely, are added to bush baths and used to make teas that are taken to treat headaches, coughs and colds, digestive and nervous problems, fever, rheumatism, cramps and so on.

Penny Royal (*Clinopodium brownei* syn. *Satureja brownei*)
Other name West Indian Thyme

The creeping, rooting stems of Penny Royal ensure that these plants develop into green mats or colonies of scented foliage. Some branches may grow to 4 in (10 cm) or even double that height but, as a rule, this is a prostrate species. However, if suspended in hanging containers, the stems will elongate indefinitely to form elegant skeins of greenery. The bruised foliage has a minty fragrance that rapidly sours to the rank, muddy undertones of freshwater plants. Penny Royal herbs, common to Tropical America and some of the larger Caribbean islands, are grown sparingly here in gardens and in pots or else they live wild in the damper regions of mainly upland districts. The plants bear small, stalked, heart-shaped leaves with irregular margins as well as scatterings of tiny, pink, tubular flowers that have purple dots.

Penny Royal plants

Penny Royal flowers

Teas brewed from a handful of green or dried plant parts are often imbibed as pleasant beverages or popular first aid, but they are mainly taken to ease bronchial problems, stomach cramps, diarrhoea, nausea, flatulence and other digestive upsets. Root decoctions serve to treat venereal disease, regulate menstrual flow and relieve chest congestion. The macerated herb is made into poultices for the chest and head to help loosen catarrh. Plant extracts are also anti-bacterial and insect repellant. Penny Royal extracts – and the essential oil in particular – contain significant amounts of pulegone, a toxic monoterpene compound that helps dissolve excess mucus in the respiratory passages, but can also cause convulsions. Hence the herb is a well-established abortifacient. Infusions made by boiling Penny Royal with Marigold, Cerasee (and a rusty nail!) are used to procure abortions. Therefore, pregnant women as well as persons suffering kidney diseases are advised not to use the herb.

This common name was adopted from the similar creeping European Penny Royal *Mentha pulegium*, a mint named for its use in repelling fleas, (*Pulex* spp.). Intriguingly, Penny Royal itself is not a description of the King's coins but has its roots in *Pulegium regium*, the name originally used by the old herbalists.

Yerba Buena – *Clinopodium douglasii* (syn. *Satureja douglasii*) – is a closely-related vine-like plant long favoured by Californians as a source for making multipurpose herbal teas.

Peppermint *(Satureja viminea)*
Other names Small Leaf Peppermint, Fine Leaf Peppermint, Wild Mint, All Heal, Savory

Peppermint bush

Peppermint flower

This is a large shrub that can grow to 9 ft (3 m) tall and it carries many airy, brittle, woody branches thickly clothed with small, glossy, apple-green leaves that have an intense peppermint fragrance. The plants are fairly common in our eastern parishes where they grow together with other wild, odourless varieties. All will bloom freely when facing the sun. They produce tiny, white or bluish flowers. This Small Leaf Peppermint is the most popular, indigenous mint. The plants are widely cultivated and are sometimes known as Bush Mint and Mint Tree. Teas made from fresh or dry material are warming, soothing and restorative, taken mainly for digestive purposes and to counteract insomnia. Mint tea with ginger is the standard remedy for stomach colic, and other infusions mixed with cayenne pepper are drunk to break up mucous congestion. Peppermint baths are taken to relieve stress, aches and pains. As the essential oil extracted from the leaves contains 35 per cent pulegone, Peppermint infusions should be taken with care.

Members of the genus *Satureja* (Savory) are characteristically small, flavoursome, mint-like plants from Europe. Most famous are Summer Savory (*S. hortensis*) and Winter Savory (*S. montana*), culinary herbs used to spice seafood, meat and legume (bean) dishes.

♦ SOLANACEAE *Tomato and Nightshade family*

The Solanaceae is an exceptionally large botanical family. It contains important food plants like aubergines, peppers, potatoes and tomatoes and other very poisonous species, *e.g.* the European Deadly Nightshade, some of which provide pesticides and drugs like atropine, nicotine and scopolamine.

Cockroach Poison *(Solanum capsicoides* syn. *S. ciliatum)*
Other name Duppy Tomato

Cockroach Poison leaves and flower

Some of these branched, leafy Cockroach Poison plants stand erect up to 3 ft (1 m) tall but the shrubs are more often seen as near prostrate growths in pastures, verges and other semi wild places in rural Jamaica. They have pinnately lobed leaves, small, white, five-part flowers, fierce prickles and an overall hairiness. There is also a purple tinge to parts of these plants: the young petioles and main veins, the undersides of flowers in bud and the ripening fruit. When ripe, the fruit turn over-bright orange or red.

Cockroach Poison fruit

Wild Susumber flowers – stalks bearing aphids and attendant ants

Cockroach Poison shrubs are neither tea bushes nor culinary herbs but poisonous plants once valued by country folk for the insecticidal properties of their marble-sized, tomato-like fruit. Opened fruit are still placed as bait in closets, drawers and other strategic positions to kill – or at least discourage – cockroaches and other insect pests. The plant poisons – bitter, water soluble, steroidal alkaloids like tomatine and solanine – are also quite toxic to humans.

Common, closely-related native Solanaceae species here include Bachelor's Pear (*Solanum mammosum*), Black Nightshade (*S. americanum*), Canker Berry (*S. bahamensis*), Susumber or Gully Bean (*S. torvum*), Wild Susumber (*S. erianthum*), Bird Pepper (*Capsicum* spp.) and the small, Wild Tomato (*Lycopersicon esculentum* var. *galena*). Susumber leaves are boiled to make muscular remedies for colds and its bitter berries and the sour Wild Tomato fruits are cooked and eaten here. The downy leaves of Wild Susumber – known as Salve Bush in The Bahamas and the Turks and Caicos Islands – are used there to make lotions to treat skin irritations.

Wild Susumber fruit and leaves

Bird Pepper *(Capsicum annuum glabriusculum* and *C. baccatum* var. *baccatum)*

Bird peppers – both Tropical American species – are our doorstep condiments. The plants can develop into semi-woody shrubs 5–6 ft (1.6–2 m) tall but, long before that stage, they go into production, bearing fiery, inch long peppers (*C. annuum glabriusculum*) or the smaller, rounder fruit of *C.*

Bird Pepper flower and green fruit

baccatum var. *baccatum* that are coveted by birds and householders alike. The feathered consumers – mockingbirds in particular – swallow the peppers whole and then go about their business sowing seed far and wide. Horticultural care extends to not pulling up the seedlings by mistake.

Bird Peppers

Jamaican cooks delight in thrusting ripe, orange red peppers – or the equally potent green ones – into marinades or any pot bubbling with savoury stew or fry. The recipients of these meals profess to enjoy the burning heat and near numbness of the mouth (anaesthesia) that is the inevitable consequence of eating fresh Bird Peppers. Although mainly used in cooking, Bird Pepper fruit and leaves are also employed in making poultices and remedies for treating asthma, various skin and mouth disorders, digestive upsets and high blood pressure. The chief bioactive substance in capsicums (including cayenne, chilli) is capsaicin, a phenolic amide well known to spread warmth and relieve pain.

♦ GESNERIACEAE *African Violet and Gloxinia family*

Search-mi-Heart *(Rytidophyllum tomentosum* syn. *Gesneria tomentosa)*

Search-mi-Heart plant in flower

Most commonly found at mid-altitudes this endemic shrub carves an existence from thin, crumbly soils on hillsides, roadside banks and on old brick and stone buildings preferably near to water. Search-mi-Heart plants actively prefer this sort of shady habitat and do not transplant easily to the relatively deep, fertile soils of cultivated gardens. The shrubs normally grow 3–6 ft (1–2 m) tall, most of which is a sprawl of bare, woody stems. Foliage is confined to the tips of branches where the long, intensely hairy, spear-shaped leaves are arranged in tight tufts. Even the leaf midribs, short petioles and young, leaf-bearing sections of stems are matted with hair. The finely-toothed, leathery leaves are velvety green on their upper surfaces but their paler undersides are heavily veined and sticky. When crushed, the aroma of mixed lemon and mint is unmistakable. Inflorescences are branched and prominently displayed at the ends of erect, 6 in (15 cm) stalks as clustered quires of gaping, tubular flowers.

Search-mi-Heart 'Hallelujah' flowers

Fresh leaves have been strewn under pillows to soothe sick children but this herb is mainly boiled down to make medicines for heart and stomach ailments. An infusion of Search-mi-Heart is the traditional treatment for a 'flittering' heart or palpitations, though the teas are also taken as morning beverages and as prescriptions for asthma, colds, coughs, congestion, stomachache and pain relief. Search-mi-Heart is sometimes used in combination with Jack-in-the-Bush (*Chromolaena odorata*) and John Charles (*Cordia globosa*) for all the above-mentioned contingencies and gout. The herb is one of many ingredients in some bottled, herbal drinks.

Cow's Tongue is the common name of the much larger *Rytidophyllum grande*, a species that has two endemic varieties.

◆ **ACANTHACEAE** *Acanthus and Shrimp Plant family*

Rice Bitters *(Andrographis paniculata)*
Other names Coolie Bitters, Wild Rice

A thicket of Rice Bitters plants in flower

This is an airy, branching herb commonly found in damp situations in the hills of eastern Jamaica. In sheltered places, the plants may form sizeable, 3 ft (1 m) tall thickets. Their square, ridged stems carry soft, dark green, paired leaves of exceedingly bitter taste, and shoulder, on slender side branches, the erect, shuttle-shaped buds and pods like rows of upended ordinances. The blooms themselves – small, white, purple-splotched, tubular structures – are usually thinly scattered about the flowering panicles. All these delicate looking flowers have wide gapes from which protrude hairy, red tipped stigmas.

Andrographis paniculata plants under many different local names are ancient medicinal herbs of the East. Indigenous to many Asian cultures, the plants are now cultivated in China and Thailand amongst other places. Immigrants from India – persons who particularly valued the bitter extracts as quinine substitutes – may have introduced the herbs to the Caribbean some time in the past. They have now spread to many other tropical and subtropical countries. The herb and its products are traded internationally and widely used in herbal mixtures.

Rice Bitters flower

Rice Bitters stem showing upright flowers and pods

The intensely bitter taste of plants like Rice Bitters, Cerasee and Neem have everywhere prompted their use in potions, stomachics, medicaments and cures. This is very much so in Jamaica where extracts of *A. paniculata*, were historically added to tonics and the strongly purgative, 'washout' mixtures given to school-age children. Beverages and bush teas made from the green parts of Rice Bitters are frequently taken to treat illnesses such as colds, fevers, diabetes, diarrhoea, dysentery, constipation and high blood pressure. The drinks are sometimes used as panaceas. Other infusions serve to bathe skin ulcers (sores). Packets of the dried, chopped herb are sold in the shops labelled as remedies for diabetes and for use in purifying the blood. But, as cuttings (and seeds) readily transplant to the garden, one may easily establish a regular supply of self-sown seedlings and fresh herb.

Extensive studies on *A. paniculata* have determined that its main medicinal compounds – bitter diterpenoids named andrographolides – boost the immune system and act strongly against many of the diseases that are treated by herbal preparations of the plant. They include those listed above as well as cancer, inflammation, kidney stones and respiratory infections. Plant extracts are also said to be analgesic and sedative.

By all accounts Rice Bitters is essentially non-toxic and safe. However, there is an anti-fertility component to the plant and pregnant women – as well as those who plan to be pregnant – are advised not to use the herb.

Fresh Cut *(Justicia pectoralis)*
Other name Carpenter's Weed

A potted plant of Fresh Cut

The name Fresh Cut may refer to three different plants, two Acanths and a Melastome. This one – *Justicia pectoralis* – familiar as domestic herbal first aid, is common to other West Indian islands, Mexico and tropical South America. The plants grow naturally here in select moist locations forming shrubby herbs. However, as they root from the nodes during growth, new plants are easily obtained from stem cuttings. These will thrive in containers or in suitable places in the open garden. Their normally green, grass-like foliage often becomes purple-tinged when older. Fresh Cut plants flower sparsely from January to May but only a few of the exotic-looking buds, borne as terminal spikes on the longer stems, ever bloom together. They make diminutive, tubular, two-lipped flowers that are essentially white but deep flushed with pink or lavender. The flowering stems of the wild grown plants may be over 2 ft (60 cm) tall.

Fresh Cut flowers

Fresh Cut is reserved mainly for the treatment of wounds and general chesty complaints, though the faintly bitter, vanilla-scented sprigs have been found in packaged salad greens. The young growth is pounded, chewed or worked between the fingers to form a sticky mass that is applied directly as a poultice onto bruises, cuts, sores and other skin injuries. These preparations are expected to stop the bleeding and close minor wounds swiftly. Rum is sometimes added to the dressings. Teas made from the leaves are taken for coughs, colds, influenza, fever, hypertension and stomachache and these potions may also be given to infants with colic. The brew made by boiling Fresh Cut leaves with Love Bush or Dodder (*Cuscuta* sp.) and an orange was thought to cure consumption (tuberculosis). Decoctions of this plant are also pressed into service to help lessen arthritis and high blood pressure and can be added to the roster of aphrodisiacs. Plant extracts are known to contain significant quantities of phenylpropanoids like coumarin and umbelliferone as well as other anti-microbial and fungicidal compounds. These substances lessen inflammation, hasten the healing of wounds and aid the relaxation of smooth body muscle.

The European Carpenter's Weed is the common Yarrow (*Achillea millefolium*), a species credited with providing immediate healing for cuts inflicted by sharp tools and relief from colds and rheumatism.

◆ **CAPRIFOLIACEAE** *Honeysuckle family*

Elder *(Sambucus nigra ssp. canadensis* syn. *S. simpsonii)*
Other names Elder Flower, Sweet Elder

Flowering Elder plant in an urban garden

The Elder plants of North America and the Caribbean are now classified as a sub-species of the European *S. nigra.* The Jamaican populations are not widely distributed and they rarely, if ever, fruit. These attractive shrubs are seen occasionally from the roadsides in hilly areas of parishes like St Andrew and St Ann and are regarded as garden escapees in the West Indies.

Elder can grow into suckering thickets of short bushes or become small trees. Flowering plants are distinctive because they carry many large heads of tiny, pure white, scented florets that are so vivid against the background of bright green foliage. The pointed, pinnate leaflets all have serrated edges.

Elderflower tea, the chief herbal remedy here, is made from fresh or dried blossoms and taken specifically for the relief of coughs, colds and influenza. But at need, teas and other plant infusions are used for blood, diabetic, digestive and venereal diseases. Elder leaves may be added to baths and lightly processed into lotions for troubled skin. A complex of chemical compounds in flowers and fruit include several flavonols (rutin, isoquercitrin), organic acids like chlorogenic acid, tannins and anthocyanins. Substances like these have known anti-inflammatory, anti-septic, anti-viral, anti-oxidant, diuretic, laxative, sedative and other beneficial properties which seem to support some of the therapeutic claims made for the plant.

Elderberries, UK

The European species – *S. nigra* and the dwarf *S. ebulus* – are both prolific wayside weeds whose freshly cut shoots possess a somewhat unpleasant odour. The shrubs bear huge crops of small juicy, purple fruit called Elderberries. These fruit – and all other plant parts – are employed in making herbal teas, infusions, drinks, *e.g.* Elderflower champagne and skin ointments. Elderberry jams jellies, pies, cordials and wines are famous. The intense dark red pigment of the berries is used as a natural food colouring. Jamaican involvement with its local Elder becomes incidental when viewed against the extensive worldwide use of, and trade in, Elder products.

♦ **RUBIACEAE** *Coffee and Madder family*

Noni *(Morinda citrifolia)*
Other names Duppy Soursop, Hog Apple, Duckfeed

Noni shrub growing on beach sand

Noni plants, native to India, the Pacific Islands and Australia, have been introduced, naturalised and cultivated here for hundreds of years. *Morinda citrifolia* plants make handsome shrubs and small trees, being well covered with bold, glossy green foliage and many small, starry white blooms. The large Noni leaves, like those of Breadfruit, Cowfoot and Oil Nut, are bound onto foreheads, stiff joints and sore places to relieve rheumatic aches and other pains. The leaves are sometimes heated before use.

Noni flower and ant sipping nectar

The fruit, however, roughly the size of big Irish potatoes, are the items of overriding herbal interest. They are oddly-shaped, composite structures, made from the fused ovaries of the tiny white flowers, and bear the external markings and 'eyes' of fruit like pineapples and soursops. Noni fruit ripen from hard green to a sickly soft, translucent yellow with an accompanying sickening smell. They are eaten raw or cooked, green or ripe in their countries of origin but elsewhere it is the fruit extracts that are important. Fermented, ripe fruit lose much of their awful odour and are infused in cold water or juiced. These extracts are drunk locally, usually as vitamin supplements, and as treatment for prostate cancer.

Noni fruit and leaves

Noni fruit are a source of vitamin C as well as containing anthraquinone compounds such as morindin. It has been suggested that the fruit also contain large quantities of a compound named proxeronine, a precursor of xeronine. This is an alkaloid chemical believed to be generated in the small intestine from its precursor. Xeronine – active in vanishingly small concentrations – is credited with helping to maintain health and well-being by regulating

Strongback flowers, immature fruit and plant bug on leaves

cell functions such as absorption of nutrients and cell repair. Noni fruit extracts are analgesic, anti-microbial, immunostimulatory and anti-tumour: properties that help to explain the widespread therapeutic applications of this herb. Extracts are taken worldwide to treat allergies, arthritis, digestive, pulmonary, skin and urinary problems, cancer, depression, fever,

Strongback fruit – Curacao

high blood pressure and many other ailments. Noni plantations abroad supply the juice concentrates and related mixtures for the lucrative commercial trade.

Strongback (*Morinda royoc*) is also known as Red Gal, Duppy Poison, Stiff Cock and Yellow Ginger. Some of these names seem to be rather strong epithets for this inoffensive-looking relation, a plant that is quite unlike the Noni shrub. Strongback plants have flexible, scrambling, vine-like branches and slender, pointed leaves. Only their flowers and relatively tiny fruit (sometimes called Goatcorn) – which are not more than 1 inch (2.5 cm) long – are visibly Noni-related. Herbal use is restricted to the roots. Freshly dug, they are a bright orange-yellow internally and are essential ingredients of the many aphrodisiac 'roots' tonic drinks. I have not yet discovered the chemical complement of this popular herb.

Strongback vine and cut roots

♦ **ASTERACEAE (COMPOSITAE)** *Daisy and Spanish Needle family*

Jack-in-the-Bush *(Chromolaena odorata* syn. *Eupatorium odoratum)*
Other names Jackney, Christmas Bush, Christmas Rose, Archangel

Jackney plant in flower

Jackney blooms from winter to spring when plants over a year old become covered with fragrant, short-lived frosty-white flower heads. (The florets of some variants are light blue to lavender in colour). Mature blooms sprout many long, white threads (stigmas) before fading. For the rest of the year, the plants exist as sprawling, woody shrubs bearing distinctive foliage. Their triangular, pointed leaves are toothed, three-veined and produce an aromatic scent when bruised. Chewed raw, the leaves provide a pleasantly stimulating taste that is only faintly bitter.

This is a high-ranking bush, one of the old pillars of Jamaican herb lore. Jackney is common in the wild – roadsides, pastures and hillsides – but the herb is also grown in backyards. All parts of these plants are worthy, but it is the young greenery that is most useful. Plant parts are crushed, applied to cuts, added to poultices and are major ingredients in bush baths. Teas brewed from the leaves are taken for bronchitis (children), colds, coughs, fever, influenza, stomach and kidney problems. Other infusions made from the flowers are also used to stem coughs as well as to treat diabetes. Tonics are made from a mixture of Jackney and other roots. Plant extracts – like those of other family members – are known to contain bitter sesquiterpene compounds that have anti-inflammatory and anti-microbial properties. The remaining constituents include organic acids, alcohols and flavones among many other phytochemicals. Two other very similar species are also common in the countryside. They are the endemic Old Woman's Bitter Bush (*Eupatorium triste*) and Bitter Bush (*Koanophyllon villosum*) whose leaves contain an exceedingly bitter substance.

Chromolaena odorata, native to Tropical America and the West Indies has now been introduced and naturalised in almost all of the remaining tropics. Unfortunately, these weeds have become seriously invasive in parts of Africa, Australia and Hawaii and other countries, where only white-tailed deer – not cattle – can be relied upon to crop the plants.

Jackney flowering shoot

Jackney – mature flowers

Herbal Plants of Jamaica

Quaco Bush *(Mikania micrantha)*

Quaco leaves

Quaco Bush is a determined climbing plant that can easily make it to the top of utility poles! It is a Tropical American vine of damp places that roots as it runs. One of nine Mikanias in Jamaica – of which five are endemic – Quaco is rampant and potentially invasive in a small garden.

The soft foliage resembles broad, ornate arrowheads. All green parts of this plant are soaked and pulped in water; usually together with the lobed

Quaco vine in flower

Quaco flowers

leaves of another climber, Cerasee, to make a bitter, sudsy green extract. The mixture is added to baths as a treatment for a range of external disorders as well as to soothe and pamper undamaged skin. Quaco juice is taken as a remedy for colic and the bush teas for colds and diarrhoea.

The vines flower freely around Christmas time producing a mass of small, fragrant, greenish-white blossom that is a seasonal decoration in its own right. The name Quaco – Guaco or Kwaku – may be of African origin possibly the name of a native healer.

Minor herbs

Following is a short, partly illustrated compilation of 24 species that have occasional, unexpected, potential or related herbal uses. Each plant is introduced by both common and scientific names. Most of the plants are gathered into interest groups, then these groupings, and the single entries, presented largely at random. For further identification, Table 1 at the end of this section lists these minor herbs in order of their botanical families.

Leaves of fruit trees

Some herbs, like the leaves of **Ackee**, **Breadfruit**, **Cashew**, **Mango** and **Tamarind** are culled from plants that already serve as indispensable food and fruit trees. These five introduced species flourish in the wild and in cultivation.

Ackee (*Blighia sapida*) the savoury contents of the red-tinged fruit pods make one half of the national dish, Ackee and Salt Fish. Ackee leaf teas are taken to address chest problems and with added salt used as a medicinal mouthwash.

Ackee fruits and leaves

Breadfruit (*Artocarpus altilis*) has cannon ball-sized fruits that make starchy staples. Breadfruit's magnificent lobed leaves are wrapped around body parts to relieve aches and pains. These leaves are also brewed into teas taken to treat high blood pressure and diabetes. The copious, sticky plant sap is applied to 'liver spots' and other skin ailments.

Breadfruits and leaves

Cashew (*Anacardium occidentale*) the roasted seeds (nuts) make tasty, popular and nutritious snacks. Cashew leaf and bark teas are taken help to control digestive disorders while the acrid seed oil supposedly banishes corns, freckles and warts.

Cashew leaves, flowers and immature fruits

Mango (*Mangifera indica*) The proliferation of trees in the wild puts the luscious, ripe, vitamin-rich fruit, mangoes, within everyone's grasp. Leaf teas are taken to treat many complaints including diarrhoea and fever while infusions made with the skins of mangoes are applied to skin sores.

Tamarind (*Tamarindus indica*) the sour-sweet flesh of the fruit pods makes beverages, snacks and spice. Extracts of the feathery leaves have multiple uses but are most often added to baths to soothe children suffering fever, chicken pox, measles and other skin irritations.

Tamarind pods and pinnate leaves

Trumpet tree (*Cecropia schreberiana*) is a very common weed tree, 60 ft (20 m) tall at maturity, whose leaf teas are taken to relieve chest and throat troubles. These infusions are also given to new mothers to help expel the placenta. As a bonus, the young buds make a good cooked vegetable. The species is also known as Wild Pawpaw, Snakewood and Guarina.

Herbal miscellany

The leafy **Comfrey** (*Symphytum officinale*) – also called **Healing Herb**, **Knit Bone** and **Bone Set** – is a relatively new arrival in Jamaica. Since then, the plant has lived up to its reputation and is also credited here with the swift healing of wounds and fractures. Comfrey is used in lotions and poultices and consumed in teas, though abroad, in its native north-temperate home, this popular herb is being restricted to external use only because of its toxic effect on the liver.

Dodder or **Love Bush** (*Cuscuta Americana*) materialises as an arresting display of orange-yellow filaments that brightens the green of trees or gilds the tops of hedges. Bits and pieces of these brittle stems are either brewed into beverages that are destined for ailing infants (marasmus) or scattered about as love charms in children's play. However, this indigenous vine is a leafless, parasitic climber that eventually weakens and kills its host plants.

Orange, spaghetti-like strands of Dodder

Lignum vitae (*Guaiacum officinale*) is a native tree famous for bearing the lavender-blue national flowers and for producing very hard, non-buoyant, resin-filled wood. Herbal use – now largely discontinued – includes drinking bitter, digestive decoctions made from wood chips; smearing the gum on skin bruises, boils and swellings and making use of the detergent and insect repellant properties of the leaves.

Lignum vitae – peeling, mosaic bark

Lignum vitae – starry blooms and shiny, paired leaflets

The many fine hairs that cover the broad, shapely, toothed leaves; flower and leaf stalks of the common **Nettle** (*Laportea aestuans*) seem non-irritant to my touch. Nettle leaf tea is taken here as a diuretic for the common relief of rheumatic and urinary problems. The plant, sometimes called White Nettle, has been dubbed 'Mortification Weed' by at least one observant herbs' person.

Nettle leaves and flowers

Exotic 'S'-shaped buds and heart-shaped leaves of *Aristolochia grandiflora*

Species that are herbal surprises to me include the beautiful **Maiden Hair Fern** (*Adiantum tenerum*) and **Country Elbow** (*Aristolochia trilobata*), one of a genus of similar-looking vines that all contain toxic aristolochic acids. The mistletoe look-alike, dubbed **Thistle Toe** (*Rhipsalis baccifera*) (which is really a cactus) and the unappealing, red-coloured cones of **John Crow Nose** or **Standing Buddy** (*Scybalium jamaicense*) are both parasitic species that live respectively on the branches and the roots of large trees. In Maroon Portland these four plants are brewed up in mixtures with several other herbs to produce tonics and medicines prescribed for complaints such as diabetes and gallstones.

Thistle Toe cactus

The yellow, daisy flowered, true **Dandelion** (*Taraxacum officinale*) and the plainer **English Plantain** (*Plantago major*), two decidedly foreign weeds, thrive here at the higher altitudes and are incorporated into local remedies for both internal and external use.

Even the weedy **Spanish Needle** (*Bidens* spp.), the touch-sensitive **Shame-o-Lady** (*Mimosa pudica*), the explosive **Duppy Gun** (*Ruellia tuberose*), the **Sea Onion** (bulb of the exquisitely flowered **Spider Lily** (*Hymenocallis latifolia*)), the hairy leaves and orange-yellow, daisy flowers of the carpeting pest plant **Mary Ghoule**, **Mary Gold**, **Marigold** or **Lady's Friend** (*Sphagneticola trilobata*) are part of the wider complement of Jamaican herbal plants.

John Crow Nose fruiting cones

Coolie Plum foliage and unripe fruits

Peppergrass in fruit

The fruits, leaves and bark of *Zizyphus* species are important herbal items in parts of Africa and China. However, the plentiful Jamaican counterpart *Z. mauritiana* known as **Coolie Plum**, **Crab Apple** and **Jujube** seems to provide only tasty snacks for children and wildlife. A second omission concerns **Wild Peppergrass** (*Lepidium virginicum*) – a member of the cabbage family that is found afoot in most places. These succulent little weeds are ignored by locals, though, I have found that washed and eaten – the entire plant; leaves, tiny white flowers and small, flattened, notched fruit – adds the tangy, curative taste of mild watercress to salads. *Lepidium sativum* (land or garden cress) is the 'cress' of those 'mustard and cress' salad sprouts. A very similar relation **Shepherd's Purse** (*Capsella bursa-pastoris*) is an abundant weed of temperate places. Shepherd's Purse was once eaten as winter greens in Britain and is traditionally used to stem excessive bleeding, diarrhoea and cystitis.

Table 1 Minor herbs

Family	Common Name	Scientific Name
ADIANTACEAE	Maiden Hair Fern	*Adiantum tenerum*
AMARYLLIDACEAE	Sea Onion (Spider Lily)	*Hymenocallis latifolia*
MORACEAE	Breadfruit Trumpet Tree	*Artocarpus altilis* *Cecropia schreberiana*
URTICACEAE	Nettle	*Laportea aestuans*
BALANOPHORACEAE	John Crow Nose (Standing Buddy)	*Scybalium jamaicense*
ARISTOLACHIACEAE	Country Elbow	*Aristolochia species*
CACTACEAE	Thistle Toe	*Rhipsalis baccifera*
BRASSICACEAE	Wild Peppergrass Cress Shepherd's Purse	*Lepidium virginicum* *L. sativum* *Capsella bursa-pastoris*
CAESALPINIACEAE	Tamarind	*Tamarindus indica*
MIMOSACEAE	Shame-o-Lady	*Mimosa pudica*
ZYGOPHYLLACEAE	Lignum vitae	*Guaicum officinale*
ANACARDIACEAE	Cashew	*Anacardium occidentale*
	Mango	*Mangifera indica*
SAPINDACEAE	Ackee	*Blighia sapida*
RHAMNACEAE	Coolie plum	*Ziziphus mauritiana*
CONVOLVULACEAE	Dodder	*Cuscuta americana*
BORAGINACEAE	Comfrey	*Symphytum officinale*
ACANTHACEAE	Duppy Gun	*Ruellia tuberosa*
PLANTAGINACEAE	English Plantain	*Plantago major*
ASTERACEAE	Dandelion Spanish Needle Mary Ghoule (Marigold)	*Taraxacum officinale* *Bidens species* *Sphagneticola trilobata*

Table 2 Plants commonly used in roots tonics

All Man Strength
Bananas (young fruit) or Banana Tree Root (*Musa* species)
Bastard Cedar (*Guazuma ulmifolia*)
Birchbark – Birch Gum, Red Birch, Balsam or Tourist Tree – (*Bursera simaruba*)
Bizzy (*Cola acuminata*)
Blood Wis
Breadnut (*Brosimum alicastrum*)
Bryal Wis – Bridal, Brial Wis or 4-Stem Herb – (*Smilax* or *Tetracera* species)
Chainy Root (*Smilax balbisiana*)
Cherry (*Malpighia* species)
Chew Stick (*Gouania acuminata*)
Coconut (young fruit) or Coconut Tree Root (*Cocos nucifera*)
Dandelion (*Senna occidentalis*)
Devil's Horse Whip (*Achyranthes aspera* var. *aspera*)
Four Man Strength (*Stemodia maritima*)
Giant Wis
Green Wis
Iron Weed or Packy Weed (*Pseudoelephantopus spicata*)
Jack-in-the-Bush (*Chromolaena odorata*)
Junction Root or Wild Coco (*Anthurium grandifolium*)
Man Back
Medina (*Alysicarpus vaginalis*)
Milk Wis (*Fosteronia floribunda*)
Nerves Wis
Pepper Elder herbs (*Peperomia* or *Piper* species)
Pimento (*Pimenta dioica*)
Puron or Pruan Tree (*Prunus occidentalis*)
Ramoon (*Trophis racemosa*)
Red sage (*Lantana camara*)
Sarsaparilla Roots (*Smilax regelii*)
Scorn-the-Earth or Mistletoe (*Oryctanthus occidentalis*)
Snake Wis (*Cissus sicyoides*)
Strongback or Golden Seal (*Morinda royac*)
Water Wis (*Vitis tilifolia*)
Wood Root (*Picrasma excelsa*)

Glossary

Note that I have enclosed the most advanced, less easily explained chemical terms in square brackets and suggest that interested readers consult some of the many, readily available, scientific texts on the matter.

Abortifacient – causes abortion.

Agar – a seaweed-derived polysaccharide or gum.

Alkaloids – a huge category of mainly colourless, alkaline, nitrogen containing plant compounds several of which make potent drugs.

Aloin – a bitter, yellow, laxative substance [an anthrone].

Amino acids – nitrogen containing compounds, 20 of which are the essential building blocks of all proteins. An additional 250 non-protein amino acids found in plants have other functions.

Analgesic – relieves pain.

Andrographolide – a very bitter substance [diterpene lactone] whose major properties include being liver-protective and stimulating the immune system.

Anethole – active ingredient [phenolic ether] present in essential oils of aniseed, fennel and other herbs and spices.

Anthelmintic or **Anti-helminthic** – destroys or expels intestinal worms.

Anthocyanins – typically blue to purple coloured compounds found in flowers and fruits, often with anti-bacterial, anti-oxidant and tissue-protective properties.

Anthraquinones and **anthrones** – closely related chemical compounds that have strongly laxative properties.

Anti-depressant – relieves depression.

Anti-dermatitis – prevents or alleviates inflammation of the skin.

Antidote – a substance or agent that can counteract adverse effects such as poisoning.

Anti-haemorrhagic – prevents or controls excessive bleeding.

Anti-inflammatory – reduces inflammation.

Anti-leukaemic – prevents or reduces the ravages caused by having too many white blood cells in body fluids.

Anti-microbial – a substance that kills or inhibits the spread of harmful microbes or germs.

Anti-oxidant – protects against cell damage by oxygen, its derivatives and other highly reactive particles.

Antiseptic – prevents or treats infection.

Anti-spasmodic – alleviates muscle contractions that cause spasms or cramps.

Anti-tumour – prevents or reduces the growth of tumours.

Anti-tussive – reduces coughing.

Anti-viral – kills or inhibits the spread of viruses.

Aphrodisiac – compound stimulating sexual desire.

Apiole – active ingredient [phenolic ether] present in the essential oils of parsley, dill and other herbs and spices.

Aristolochic acids – toxic [aporphine] alkaloids that can interfere with DNA, cause cancers and kidney damage.

Aromatherapy – the use of fragrant plant products such as essential oils to improve health and well being.

Aromatic – produces a noticeable aroma or smell.

Ascaridole – a highly toxic compound [monoterpenoid peroxide] that kills or expels intestinal parasites such as worms.

Aspirin – [acetylsalicylic acid] a popular, synthetic, pain-relieving drug.

Astringent – protects by coating and contracting tissues such as those on the surfaces of wounds and swollen mucous membranes.

Atropine – a potent [tropane] alkaloid drug that has many properties including stimulating the CNS and relaxing smooth muscle.

Azadirachtin – bitter, triterpenoid [limonoid] compound that has strong insecticidal, anti-microbial and cosmetic properties.

Bactericidal – destroys bacteria.

Berberine – potent, yellow [protoberberine] alkaloid that is markedly antibiotic, anti-viral and digestive. It also can affect the heart rate.

Betaine – a sweet tasting, widely distributed nutrient [trimethylglycine] found in both plants and animals.

Biennial – a plant that takes two growing seasons or years to complete its life cycle.

Bile – bitter liquid made in the liver, stored in the gall bladder and used by the body to help digest fats.

Bitters – compounds which stimulate the appetite and encourage the flow of digestive juices.

Bixin – a carotenoid (orange-red) compound that helps heal wounds and burns and provides colouring for food and cosmetics.

Bract – small, leaf-like structure that usually lies directly beneath each flower or inflorescence.

Bronchorelaxant – relaxes spasms of the bronchial tubes of the lungs.

Caffeic acid – a reactive phenylpropanoid [hydroxycinnamic acid] compound found in coffee beans that is analgesic, anti-inflammatory and stimulates the gut.

Caffeine – a [purine] alkaloid found in coffee, tea and other plant species that is universally valued as a general stimulant.

Camphene – a compound [bicyclic monoterpene] with a camphor-like odour.

Camphor – strongly scented, active ingredient of the essential oil extracted from the camphor tree. The compound [a bicyclic ketone] has many functions including stimulating the CNS and circulatory systems, reducing flatulence, dissolving mucus and being warming.

Cannabinoids – a class of phenolic terpenoids the best known of which is delta-9 tetrahydrocannabinol (THC), the active principle of marijuana.

Cardiotonic – stimulates and strengthens the heart.

Carophyllene – a volatile, sesquiterpene compound with a sweet, woody, spicy scent found in several common plants.

Carotene – a member of the large, yellow to red coloured, carotenoid group of tetraterpenes. Carotene can be converted into vitamin A in the gut.

Cataract – a disease of the eye where the lens becomes progressively clouded.

Catarrh – the discharge in nose or throat produced by inflamed mucous membranes.

Catkin – a long, thin, spike-like inflorescence that generally hangs from a branch.

Chlorogenic acid – a reactive phenylpropanoid [polyphenol] compound found in high concentrations in coffee beans. The substance is strongly antioxidant and possibly antitumour.

Choleretic – a substance that boosts the production of bile.

Cholesterol – an essential fatty substance [steroid alcohol] present in the body. Heart disease has been linked to elevated cholesterol levels in the blood.

Cineole – found in very many plant species but is dominant in eucalyptus oil. The compound [a monoterpene oxide] is used commercially in remedies to dissolve mucus and control coughs.

Cinnamaldehyde – cinnamon scented, phenylpropanoid compound present in quantity in the essential oil extracted from cinnamon bark with marked anti-spasmodic properties.

Citral – an highly aromatic substance [linear monoterpene aldehyde] that provides the soothing, anti-microbial properties associated with extracts from citrus plants.

CNS – central nervous system.

Cultivar – a contraction of 'cultivated variety', a selected form of a cultivated plant.

Curcumin – a yellow phenylpropanoid [diarylheptanoid] compound. This group of chemicals exhibits anti-inflammatory, stress-reducing and liver-protective properties.

Cyanide – chemical combination of carbon and nitrogen that is invariably toxic.

Decongestant – relieves congestion by removing mucus and opening up the air passages.

Diaphoretic – increases sweating.

Dimeric – made up of two units.

Diterpenes – 20 carbon atom compounds, often very bitter-tasting.

Diuretic – promoting urination.

DNA – [deoxyribonucleic acid] complicated macromolecules that are the genetic material of cells.

Ecdysterone – a plant steroid that acts as a moulting hormone in insects and is thought to have potential as a muscle-building, growth additive for humans.

Emetic – a substance that promotes vomiting.

Endemic – a species that is native to a particular place or region.

Eugenol – a highly aromatic, phenylpropanoid alcohol, the active ingredient of oil of cloves. Eugenol is present in other spices, *e.g.* nutmeg and pimento (allspice) and is best known as a powerful anaesthetic (dentistry), anti-microbial and anti-convulsant agent.

Expectorant – a substance that promotes the production and removal of mucus from the respiratory system.

Flavonoids – universal, pigmented plant constituents [polyphenols] with a wide range of protective functions such as being anti-oxidant, antiviral, liver protective, anti-inflammatory and anti-tumour.

Fungicidal – kills or destroys fungal infections.

Geranial – a form of citral [linear, monoterpene aldehyde] that is partly responsible for the aroma and the soothing, antimicrobial properties of citrus oils and extracts.

Germplasm – genetic material.

Gingerols – a series of pungent compounds [phenolic ketones] found in the Zingiberaceae family of plants. These substances are analgesic, anti-oxidant, anti-tumour, anti-tussive, anti-ulcer and provide effective treatment for indigestion.

Glycoprotein – a compound made up of chemically joined sugar and protein molecules.

Gout – arthritic inflammation of some joints due to raised levels in the blood of the waste material, uric acid.

Gums – sticky, water soluble plant secretions, *e.g.* gum arabic from African *Acacia* trees.

HIV – human immunodeficiency virus that causes AIDS.

Hydrocarbons – compounds that contain the elements hydrogen and carbon only.

Hypertension – high blood pressure.

Hypoglycaemic – unusually low levels of blood sugar.

Hypotension – low blood pressure.

In vitro – in a laboratory environment.

Indigenous – native to a particular region; not introduced.

Inflammation – response to damage or infection that is characterised by swelling, redness and usually pain.

Lectins – Carbohydrate-binding proteins, often causing cell death in animal tissues. They occur widely in plants of the pea family but are rendered harmless by cooking.

Limonene – a hydrocarbon [monocyclic monoterpene] that is the major constituent of citrus oils and known to be antiviral and decongestant.

Linalool or **linanol** – a potent, widely distributed alcohol [linear monoterpene] found in rose, lavender, lemon, mint, basil and many other plants. This fragrant substance acts upon the CNS and is known to be sedative, spasmolytic and anaesthetic.

Marasmus – starvation or protein-calorie deficiency disease of children.

Menthol – an alcohol [monocyclic monoterpene] found largely in the mint family, *e.g.* peppermint (*Mentha piperita*) that is a strong spasmolytic and decongestant.

Monoterpene – volatile, 10-carbon atom compound, widespread component of essential oils.

Myrcene – a hydrocarbon [linear monoterpene] present in many common plants. It has a fresh, spicy, balsamic scent.

Neral – a form of citral.

Neutraceutical – any natural, biologically active compound with proven positive effects on human health and happiness.

Oedema – body tissues that become swollen with fluid.

Oleorescin – a mixture of essential oil and resin that can be extracted from some plants.

Organic acid – simple, sharp-tasting organic compound, like citric acid (citrus fruits), acetic acid (vinegar) and tartaric acid (tamarind fruits).

Phenol – aromatic alcohol containing the definitive (OH) group attached to a benzene ring.

Phenylpropanoid – any of a huge group of rather active compounds whose core structure is based on a benzene ring with a three carbon, hydrocarbon chain.

Phlegm – thick mucus produced in the throat and nose.

Phytosterol – fat-like, triterpenoid compound, *e.g.* sitosterol and stigmasterol, that occurs in the cells and membranes of plants. May help control cholesterol levels in the blood and may also be anti-tumour.

Pinene – a compound [bicyclic monoterpene] that is dominant in the aromatic essential oils (*e.g.* turpentine) extracted from coniferous (pine) trees.

Piperine – a pungent alkaloid substance [acid-amide] that has liver-protective properties and appears to assist the uptake of nutrients from the small intestine.

PMS – (premenstrual syndrome) a clutch of annoying symptoms that may appear shortly before each monthly period.

Podophyllotoxin – a complex, anti-cancer, phenylpropanoid compound.

Polysaccharide – polymer made up of strings of chemically linked sugar molecules, *e.g.* starch.

Psoriasis – a chronic, inherited skin disease.

Pulegone – a toxic compound [monocyclic ketone] that can cause convulsions

Purgative – causing the bowels to empty.

Resin – solid, complex plant secretion that is insoluble in water, *e.g.* cannabis resin and myrrh.

Rosemarinic acid – a phenylpropanoid compound that acts like a tannin.

Sanguinarine – a potent, red, alkaloid [protoberberine] that is strongly antimicrobial and antiviral.

Saponin – triterpenoid compound with detergent or soapy properties affecting cell membranes and often having a wide range of biological activities (*e.g.* anti-cancer, anti-inflammatory, anti-diabetic, anti-fungal, diuretic, digestive, wound-healing).

Sesquiterpene – volatile, sometimes bitter, 15-carbon atom compound; typical component of essential oils.

Sitosterol –common, fatty plant sterol or phytosterol, [a steroidal alcohol].

Soporific – calming and sleep-inducing.

Stamen – male, pollen-producing part of a flower.

Steroid – fatty, triterpene derived compound (containing 27 or fewer carbon atoms) often with hormonal or pharmacological properties.

Stigma – female, feathery pollen-catching part of a flower.

Stomachic – boosts the appetite and functioning of the digestive system.

Tannin – widely distributed, highly reactive compound [polyphenol] having astringent and anti-oxidant properties.

Terpene – an extensive group of biologically active plant compounds whose structures are based on a series of linked 5-carbon [isoprene] units. Terpenes or Terpenoids are defined by their carbon (C) atom content, being sub-classified into monoterpenes (C_{10}), sesquiterpenes (C_{15}), diterpenes (C_{20}), triterpenes (C_{30}), tetraterpenes (C_{40}), and polyterpenes ($>C_{40}$).

Tetraterpene – a class of terpene having a 40-carbon atom core, mainly orange-red pigmented carotenoid sunstances.

Therapeutic – curative or healing.

Thymol – a compound [monoterpene phenol] that is the major constituent of the essential oils of herbs like thyme and marjoram. Thymol is a potent substance that is known to be anti-microbial, expectorant, spasmolytic and a vermifuge.

Triterpene – a large and varied subsection of terpenes including steroids, saponins and phytosterols all derived from molecules that contain 30 carbon atoms.

Tuber – swollen, underground plant stem or root that functions as a food store.

Vermifuge – kills or expels worms and other intestinal parasites.

Zingiberene – volatile sesquiterpene compound found in plants like ginger and turmeric. Effective in treating indigestion and gastric ulcers.

Bibliography

Adams, C.D. (1972) *Flowering plants of Jamaica*, University of the West Indies, Mona, Jamaica.

Bailey-Shaw, Y.A., Gallimore, W.A. and Reid, C.S. (2000 and 2001) 'The analysis and applicability of Jamaican ginger oleoresins to the nutraceutical industry', *Jamaican Journal of Science and Technology*, vols 12 and 13, pp 80–92.

Harris, I., Harris, L. and Henry, L.G. (2004) *Common Medicinal Plants of Portland, Jamaica*, MAPCO Business Printers Ltd., Kingston, Jamaica.

Johnson, L.B., Williams, L.A.D. and Roberts, E.V. (1997) 'An insecticidal and acaricidal polysulphide metabolite from the roots of *Petiveria alliaceae*', *Pesticide Science*, vol 50, pp. 228–32.

Lowe, H., Payne-Jackson, A., Beckstrom-Sternberg, S.M. and Duke, J. (2000) *Jamaica's Ethnomedicine*, Canoe Press, University of the West Indies, Mona, Kingston, Jamaica.

Lowe, H., Morrison, E.Y.St.A., Magnus, K. and Campbell-Grizzle, E. (2002) *Poisonous Plants of Jamaica*, Pelican Publishers Ltd., Kingston, Jamaica.

Lowe, H., Morrison, E.Y. and Mitchell, S.A. (2004) *Caribbean Medicinal Plants*, CD e-book, Pelican Publishers Ltd., Kingston, Jamaica.

Senior, O. (2003) *Encyclopedia of Jamaican Heritage*, Twin Guinep Publishers Ltd., St Andrew, Kingston, Jamaica.

Pengelly, A. (2004) *The Constituents of Medicinal Plants*, CABI Publishing, CAB International.

Van Wyk, B-E. and Wink, M. (2004) *Medicinal Plants*, Timber Press Inc., Portland, Oregon, USA.

Index to Botanical Names

Index to Common Names